MUST KNOW HIGH SCHOOL TRIGONOMETRY

Sandra Luna McCune, PhD

William D. Clark, PhD

Mc
Graw
Hill

New York Chicago San Francisco Athens London Madrid
Mexico City Milan New Delhi Singapore Sydney Toronto

P9-CLS-757

Copyright © 2020 by McGraw-Hill Education. All rights reserved. Printed in the United States of America. Except as permitted under the United States Copyright Act of 1976, no part of this publication may be reproduced or distributed in any form or by any means, or stored in a database or retrieval system, without the prior written permission of the publisher.

1 2 3 4 5 6 7 8 9   LCR   24 23 22 21 20 19

ISBN   978-1-260-45988-3
MHID   1-260-45988-8

e-ISBN   978-1-260-45989-0
e-MHID   1-260-45989-6

Interior design by Steve Straus of Think Book Works.
Cover and letter art by Kate Rutter.

McGraw-Hill Education books are available at special quantity discounts to use as premiums and sales promotions or for use in corporate training programs. To contact a representative, please visit the Contact Us pages at www.mhprofessional.com.

Contents

Introduction   vii

The Flashcard App   ix

1   Angles and Their Measure   1

Basic Concepts and Terminology of Angles   2

Complementary and Supplementary Angles   6

Coterminal Angles   6

Reference Angles   7

Radian Measure   9

Applications of Radian Measure   11

2   Concepts from Geometry   19

The Sum of a Triangle's Angles   20

The Triangle Inequality Theorem   21

The Pythagorean Theorem   22

Similar Triangles   26

3   Right Triangle Trigonometry   34

Trigonometric Ratios of an Acute Angle in a Right Triangle   35

Trigonometric Ratios of Special Acute Angles   38

Finding Missing Sides in Right Triangles with Special
Acute Angles   40

4   The Trigonometry of General Right Triangles   45

Solving Right Triangles   46
Applications of Right Triangle Trigonometry   48

5   The Trigonometry of Oblique Triangles   53

Law of Cosines (SAS or SSS)   54
Law of Sines (ASA or AAS)   58
Law of Sines Ambiguous Case (SSA)   63
Solving General Triangles   66
Area of a General Triangle Using Trigonometry   69

6   Trigonometric Functions of Any Angle   81

Trigonometric Functions of Complementary Angles   87
Unit Circle Trigonometry   89
Trigonometric Functions of Quadrantal Angles   91
Trigonometric Functions of Coterminal Angles   94
Trigonometric Functions of Negative Angles   94
Using Reference Angles to Find the Values of Trigonometric Functions   96

7   Trigonometric Identities   106

Definition and Guidelines   107
The Reciprocal and Ratio Identities   109
The Cofunction, Periodic, and Even-Odd Identities   110

The Pythagorean Identities   112
Sum and Difference Formulas for the Sine Function   114
Sum and Difference Formulas for the Cosine Function   116
Sum and Difference Formulas for the Tangent Function   117
Double-Angle Identities   119
Half-Angle Identities   121

8 Trigonometric Functions of Real Numbers

132

Definitions and Basic Concepts of Trigonometric Functions
of Real Numbers   133
Periodic Functions   135

9 Graphs of the Sine Function

142

The Graph of $y = \sin x$ 143
The Graph of $y = A \sin x$ 146
The Graph of $y = A \sin Bx$ 148
The Graph of $y = A \sin(Bx - C)$ 150
The Graph of $y = A \sin(Bx - C) + D$ 153

10 Graphs of the Cosine Function

162

The Graph of $y = \cos x$ 163
The Graph of $y = A \cos(Bx - C) + D$ 165

11 Graphs of the Tangent Function

172

The Graph of $y = \tan x$ 173
The Graph of $y = A \tan(Bx - C) + D$ 174

12 Inverse Trigonometric Functions 177

Definitions and Basic Concepts of the Inverse Sine, Cosine, and Tangent Functions 178

Evaluating the Inverse Sine, Cosine, and Tangent Functions 182

13 Solving Trigonometric Equations 190

Solving for Exact Solutions to Trigonometric Equations 193

Solving for Approximate Solutions to Trigonometric Equations 199

Answer Key 204

Appendix A: Calculator Instructions for Trigonometry Using the TI-84 Plus 246

Appendix B: Trigonometric Identities/Formulas 255

Introduction

Welcome to your new trigonometry book! Let us try to explain why we believe you've made the right choice. This probably isn't your first rodeo with either a textbook or other kind of guide to a school subject. You've probably had your fill of books asking you to memorize lots of terms (such as in school). This book isn't going to do that—although you're welcome to memorize anything you take an interest in. You may also have found that a lot of books jump the gun and make a lot of promises about all the things you'll be able to accomplish by the time you reach the end of a given chapter. In the process, those books can make you feel as though you missed out on the building blocks that you actually need to master those goals.

With *Must Know High School Trigonometry*, we've taken a different approach. When you start a new chapter, right off the bat you will immediately see one or more **must know** ideas. These are the essential concepts behind what you are going to study, and they will form the foundation of what you will learn throughout the chapter. With these **must know** ideas, you will have what you need to hold it together as you study, and they will be your guide as you make your way through each chapter.

To build on this foundation, you will find easy-to-follow discussions of the topic at hand, accompanied by comprehensive examples that show you how to apply what you're learning to solving typical trigonometry questions. Each chapter ends with review questions—more than 400 throughout the book—designed to instill confidence as you practice your new skills.

This book has other features that will help you on this trigonometry journey of yours. It has a number of sidebars that will either provide helpful information or just serve as a quick break from your studies. The **BTW** sidebars ("by the way") point out important information, as well as tell you what to be careful about trigonometry-wise. Every once in a while, an ⊕ IRL sidebar ("in real life") will tell you what you're studying has to do with the real world; other IRLs may just be interesting factoids.

In addition, this book is accompanied by a flashcard app that will give you the ability to test yourself at any time. The app includes more than 135 "flashcards" with a review question on one "side" and the answer on the other. You can either work through the flashcards by themselves or use them alongside the book. To find out where to get the app and how to use it, go to the next section, "The Flashcard App."

Before you get started, though, let me introduce you to your guides throughout this book. Dr. Sandra L. McCune is a former Regents Professor who taught mathematics at Stephen F. Austin State University. Dr. William D. Clark is a professor of mathematics at the same university. We've had the pleasure of working with Drs. McCune and Clark before. They have a clear idea what you should get out of a trigonometry course and have developed strategies to help you get there. Drs. McCune and Clark have also seen the kinds of pitfalls that students can fall into, and they are experienced hands at solving those difficulties. In this book, they apply that experience both to showing you the most effective way to learn a given concept and how to extricate yourself from any trouble you may have gotten into. They will be trustworthy guides as you expand your trigonometry knowledge and develop new skills.

Before we leave you to Drs. McCune and Clark's sure-footed guidance, let us give you one piece of advice. While we know that saying something "is the *worst*" is a cliché, if anything *is* the worst in trigonometry, it could be learning how to verify and use trigonometric identities. Let Drs. McCune and Clark introduce you to the concept and show you how to work confidently with identities. Take our word for it, mastering trigonometry identities will leave you in good stead for the rest of your math career.

Good luck with your studies!

The Editors at McGraw-Hill

The Flashcard App

his book features a bonus flashcard app. It will help you test yourself on what you've learned as you make your way through the book (or in and out). It includes 135+ "flashcards," both "front" and "back." It gives you two options as to how to use it. You can jump right into the app and start from any point that you want. Or you can take advantage of the handy QR codes at the end of each chapter in the book; they will take you directly to the flashcards related to what you're studying at the moment.

To take advantage of this bonus feature, follow these easy steps:

Search for **Must Know High School** app from
either Google Play or the App Store.

↓

Download the app to your smartphone or tablet.

↓

Once you've got the app,
you can use it in either of two ways.

↙   ↘

Just open the app and you're ready to go.	Use your phone's QR code reader to scan any of the book's QR codes.
You can start at the beginning, or select any of the chapters listed.	You'll be taken directly to the flashcards that match your chapter of choice.

↘   ↙

Get ready to test your trigonometry knowledge!

1 Angles and Their Measure

MUST ⚡ KNOW

⚡ An angle is formed by two rays with a common endpoint and is measured in degrees or radians.

⚡ All reference angles measure between 0° and 90°.

⚡ Angular speed is the rate of change of the measure of the central angle of an object moving in a circular path.

⚡ Knowing the measure of the central angle of a circle enables us to calculate the length of an arc or area of a sector of the circle.

⚡ Angles provide the foundation for the definitions of the trigonometric functions.

T he concept of an angle is fundamental to the study of trigonometry. The basic relationships among angles that arise in the study of trigonometry are the groundwork for further topics. This chapter presents the most necessary of those relationships to know.

Basic Concepts and Terminology of Angles

What is an angle? Let's begin with the definition of a ray and then go from there. A **ray** is a half-line beginning at an endpoint and extending indefinitely in one direction from that point. An **angle** is formed by two rays with a common endpoint. The two rays are the angle's **sides**, and the common endpoint is its **vertex**. The angle begins with the two rays lying on top of one another. One ray, the **initial side**, is fixed in place, and the other ray, the **terminal side**, is rotated about the vertex. You can indicate the rotation with a small arrow close to the vertex.

Customarily, labels for angles are Greek letters, for example, α (alpha), β (beta), γ (gamma), θ (theta), and ϕ (phi), as well as uppercase and lowercase Roman letters, such as A, B, C, x, y, and z. The following figure shows the angle θ.

Angle θ

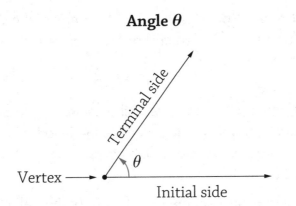

The **measure of an angle** is the amount of rotation from the initial side to the terminal side. As shown in the next figure, *counterclockwise* rotations yield positive angles, and *clockwise* rotations yield negative angles.

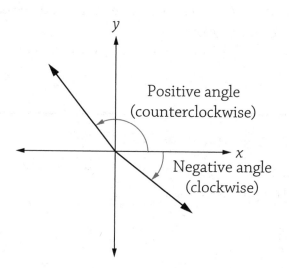

Positive angle
(counterclockwise)

Negative angle
(clockwise)

> **BTW**
>
> *The measure of an angle depends on the amount of rotation that makes the angle, not on the perceived lengths of the sides forming the angle.*

You are likely familiar with measuring angles using degrees. One **degree** (1°) is $\frac{1}{360}$ of a full revolution (counterclockwise) about the vertex. So, one complete counterclockwise revolution corresponds to 360°.

> **IRL** The system of measuring angles in degrees, such that 360° is one complete circular revolution, originated in ancient Babylonia.

> **IRL** Similar to the division of hours into minutes and seconds of time, degrees can be further divided into minutes and seconds. Each degree equals 60 **minutes** ('), and each minute equals 60 **seconds** ("). So, for instance, 40 degrees 15 minutes 30 seconds is written 40°15'30". In the real world, this scheme is not as customary as it used to be. Nowadays, parts of degrees are expressed decimally, such as 22- and one-half degrees expressed as 22.5°, rather than as 22°30'.

In the coordinate plane, an angle in **standard position** has its vertex located at the origin, and its initial side extends along the positive *x*-axis. The terminal side can rotate in either a counterclockwise or a clockwise direction and can lie in any quadrant or on any axis.

When the terminal side lies in a quadrant, we say the angle lies in that quadrant.

EXAMPLE

▶ In what quadrant does an angle that measures 135° lie?

▶ We construct this angle in the coordinate plane by rotating 135° counterclockwise from the positive *x*-axis.

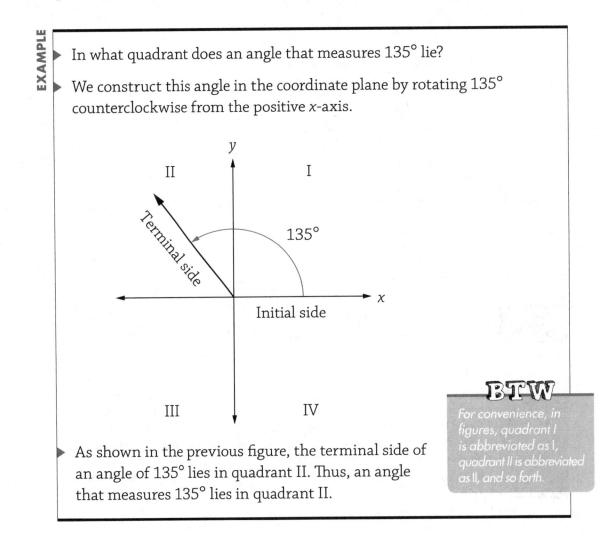

▶ As shown in the previous figure, the terminal side of an angle of 135° lies in quadrant II. Thus, an angle that measures 135° lies in quadrant II.

BTW

For convenience, in figures, quadrant I is abbreviated as I, quadrant II is abbreviated as II, and so forth.

If the terminal side of an angle lies on an axis, the angle is a **quadrantal angle**.

BTW

Recall from elementary geometry that an angle is **acute** if it is between 0° and 90°, a **right angle** if it equals 90°, an **obtuse angle** if it is between 90° and 180°, a **straight angle** if it equals 180°, and a **reflex angle** if it is between 180° and 360°.

EXAMPLE

▶ In what quadrant does an angle that measures 90° lie?

▶ We construct this angle in the coordinate plane by rotating 90° counterclockwise from the positive *x*-axis.

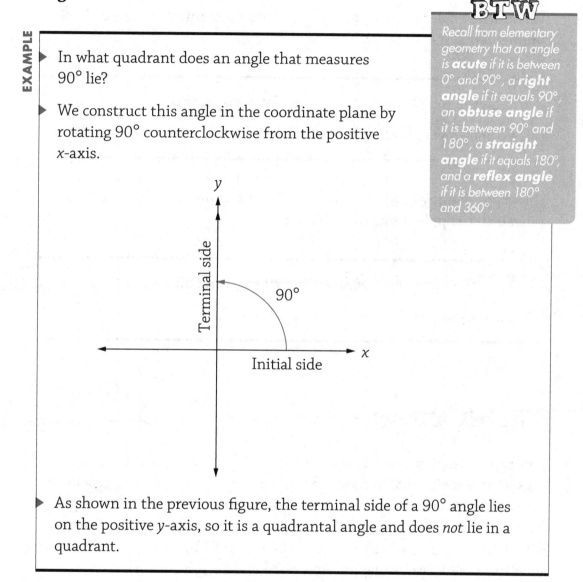

▶ As shown in the previous figure, the terminal side of a 90° angle lies on the positive *y*-axis, so it is a quadrantal angle and does *not* lie in a quadrant.

Complementary and Supplementary Angles

Two positive angles whose sum is 90° are **complementary** or, equivalently, **complements** of each other.

EXAMPLE

▸ What is the complement of an angle that measures 30°?

▸ The complement of 30° is 90° − 30° = 60°.

Two positive angles whose sum is 180° are **supplementary** or, equivalently, **supplements** of each other.

EXAMPLE

▸ What is the supplement of an angle that measures 135°?

▸ The supplement of 135° is 180° − 135° = 45°.

Coterminal Angles

Coterminal angles have the same initial and terminal sides. To find an angle that is coterminal to a given angle, add or subtract 360°. In general, a given angle θ is coterminal with $\theta + n(360°)$, where n is a nonzero integer. If an angle is greater than 360° or is negative, we can find an equivalent nonnegative coterminal angle that is less than 360° by adding or subtracting an adequate positive integer multiple of 360°.

EXAMPLE

▶ Determine an angle between $0°$ and $360°$ that is coterminal with an angle that measures $405°$.

▶ The angle $45°$ is coterminal with $405°$ because $405° - 360° = 45°$.

The process works whether the angle is positive, as in the previous example, or negative, as in the following example.

EXAMPLE

▶ Determine an angle between $0°$ and $360°$ that is coterminal with an angle that measures $-30°$.

▶ The angle $330°$ is coterminal with $-30°$ because $-30° + 360° = 330°$.

Reference Angles

The **reference angle** for a non-quadrantal angle in standard position is the acute angle formed by the terminal side of the angle and the x-axis. All reference angles have measures between $0°$ and $90°$.

The relationship of a positive angle θ that is less than $360°$ and its reference angle θ in each quadrant is given in the following table.

θ's Quadrant	θ_r Relationship
I	$\theta_r = \theta$
II	$\theta_r = 180° - \theta$
III	$\theta_r = \theta - 180°$
IV	$\theta_r = 360° - \theta$

EXAMPLE

▶ Determine the reference angle for 135°.

▶ 135° is in quadrant II, so its reference angle is $180° - 135° = 45°$.

For a non-quadrantal angle θ that is greater than 360°, we first find its positive conterminal angle that is less than 360°, then we determine the reference angle for the conterminal angle.

EXAMPLE

▶ Determine the reference angle for 930°.

▶ Because $930° > 360°$, find its coterminal angle first:
$930° - (2 \times 360°) = 930° - 720° = 210°$.

▶ Then determine the reference angle for 210°:

▶ 210° is in quadrant III, so its reference angle is $210° - 180° = 30°$.

For a non-quadrantal negative angle θ, first we find its positive conterminal angle that is less than 360°, then we determine the reference angle for the conterminal angle.

EXAMPLE

▶ Determine the reference angle for −60°.

▶ Because the angle is negative, find its coterminal angle first:
$-60° + 360° = 300°$.

▶ Then determine the reference angle for 300°:

▶ 300° is in quadrant IV, so its reference angle is $360° - 300° = 60°$.

Radian Measure

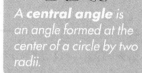

A **radian** is the measure of a central angle of a circle that intercepts an arc whose length equals the radius of that circle.

BTW
A *central angle* is an angle formed at the center of a circle by two radii.

Because the circle's circumference equals 2π times the radius, a full circular rotation is 2π radians. Therefore, we have the following equivalencies:

$$2\pi \text{ radians} = 360°$$
$$\pi \text{ radians} = 180°$$
$$1 \text{ radian} = \frac{180°}{\pi}(\approx 57.3°)$$
$$\frac{\pi}{180} \text{ radians} = 1°$$

BTW
It is not necessary to write the units "radians" after a radian measure. You can assume that an angle measure that is not labeled with "degrees" or the degree symbol is in radians.

To convert degrees to radians, we multiply the degree measure by $\frac{\pi}{180°}$. It is commonplace to express the radian measure in terms of π.

EXAMPLE

▶ Convert $60°$ to radians in terms of π.

▶ We multiply the degree measure by $\frac{\pi}{180°}$:

$$60°\left(\frac{\pi}{180°}\right) = \cancel{60°}\left(\frac{\pi}{\cancel{180°}}\right) = \frac{\pi}{3}$$

To convert radians to degrees, we multiply the radian measure by $\dfrac{180°}{\pi}$.

EXAMPLE

▶ Convert a radian measure of $\dfrac{3\pi}{4}$ to degrees.

▶ We multiply the radian measure by $\dfrac{180°}{\pi}$:

$$\frac{3\pi}{4}\left(\frac{180°}{\pi}\right)=\frac{3\cancel{\pi}}{\underset{1}{\cancel{4}}}\left(\frac{\overset{45°}{\cancel{180°}}}{\cancel{\pi}}\right)=135°$$

In the previous example, you likely observed that because the radian measure was expressed in terms of π, the π factors canceled out when we multipled by $\dfrac{180°}{\pi}$. In the next example, the radian measure is expressed numerically without the use of π. In such a case, we will round the answer as indicated.

EXAMPLE

▶ Convert a radian measure of 3 to degrees (round the answer to the nearest tenth of a degree).

▶ We multiply the radian measure by $\dfrac{180°}{\pi}$:

$$3\left(\frac{180°}{\pi}\right)=\frac{540°}{\pi}\approx171.9°$$

 IRL   In calculus and other branches of higher mathematics and in physics and other sciences, angles are almost universally measured in radians.

Applications of Radian Measure

Following are some applications of radian measure of angles.

- **Arc length**   On a circle of radius r, a central angle θ (measured in radians), intercepts an arc of length $s = r\theta$.

EXAMPLE

▶ Find the length of the arc intercepted by a central angle of $\dfrac{5\pi}{6}$ in a circle of radius 24 inches (see the following figure). (Round the answer to one decimal place.)

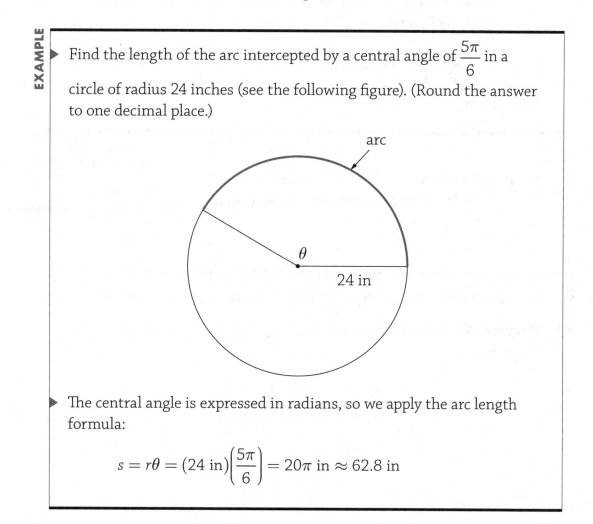

arc

θ

24 in

▶ The central angle is expressed in radians, so we apply the arc length formula:

$$s = r\theta = (24 \text{ in})\left(\frac{5\pi}{6}\right) = 20\pi \text{ in} \approx 62.8 \text{ in}$$

When finding the arc length intercepted by a central angle expressed in degrees, we first convert the degree measure to radians, then we apply the arc length formula.

EXAMPLE

▶ Find the length of the arc intercepted by a central angle of 315° in a circle of radius 36 centimeters. (Round the answer to one decimal place.)

▶ First, $315°\left(\dfrac{\pi}{180°}\right) = \dfrac{7\pi}{4}$; then $s = r\theta = (36 \text{ cm})\left(\dfrac{7\pi}{4}\right) = 63\pi$ cm

≈ 197.9 cm.

■ **Area of a sector**   The area of a sector of a circle of radius r and central angle θ (measured in radians) is $A = \dfrac{1}{2}r^2\theta$.

BTW

*A **sector** is the portion of the interior of a circle enclosed by two radii and an arc of the circle.*

EXAMPLE

▶ Find the area of a sector intercepted by a central angle of $\dfrac{5\pi}{6}$ in a circle of radius 24 inches (see the following figure). (Round the answer to one decimal place.)

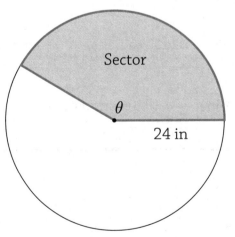

Sector

θ

24 in

▶ The central angle is expressed in radians, so we apply the formula for the area of a sector:

$$A = \frac{1}{2}r^2\theta = \frac{1}{2}(24 \text{ in})^2\left(\frac{5\pi}{6}\right) \approx 754.0 \text{ in}^2$$

When finding the area of a sector with a central angle expressed in degrees, we first convert the degree measure to radians, then we apply the formula for the area of a sector.

EXAMPLE

▶ Find the area of a sector with a central angle of 315° in a circle of radius 36 centimeters. (Round the answer to one decimal place.)

▶ First, $315°\left(\dfrac{\pi}{180°}\right) = \dfrac{7\pi}{4}$; then $A = \dfrac{1}{2}r^2\theta = \dfrac{1}{2}(36 \text{ cm})^2\left(\dfrac{7\pi}{4}\right)$

$\approx 3{,}562.6 \text{ cm}^2$.

- **Angular speed**   The angular speed of an object moving at a constant speed along a circular arc is $\dfrac{\theta}{t}$, where θ is the central angle measured in radians and t is the elapsed time.

BTW

Angular speed *is the rate at which an object turns. It is the angle through which a rotating object travels per unit of time.*

EXAMPLE

▶ An engine is rotating at 540 revolutions per minute. Calculate its angular speed in radians per second.

▶ Given that 1 revolution is 2π radians and 1 minute is 60 seconds, 540 revolutions per minute in radians per second is

$$\frac{(540)(2\pi \text{ radians})}{60 \text{ sec}} = 18\pi \text{ radians per second.}$$

EXERCISES

EXERCISE 1-1

State the quadrant in which the angle lies or the axis on which it lies.

1. 20°

2. 180°

3. 97°

4. 236°

5. −18°

EXERCISE 1-2

For questions 1 to 5, find the complement of the given angle.

1. 40°

2. 16°

3. 88°

4. 54°

5. 47°

For questions 6 to 10, find the supplement of the given angle.

6. 120°

7. 22°

8. 81°

9. 90°

10. 101°

EXERCISE 1-3

Determine an angle between 0° and 360° that is coterminal with the given angle.

1. −17°

2. 380°

3. 800°

4. −101°

5. 460°

EXERCISE 1-4

Find the reference angle for the given angle.

1. 150°

2. 225°

3. 300°

4. −17°

5. 1155°

EXERCISE 1-5

For questions 1 to 5, convert the degree measure to radians in terms of π.

1. 30°

2. 120°

3. 45°

4. 90°

5. 135°

For questions 6 to 10, convert the radian measure to degrees.

6. 3π

7. $\dfrac{5\pi}{4}$

8. $-\dfrac{\pi}{4}$

9. $\dfrac{5\pi}{6}$

10. $\dfrac{9\pi}{4}$

For questions 11 to 15, convert the radian measure to degrees. (Round answers to one decimal place, as needed.)

11. 3.5

12. 1.4

13. 2.3

14. 4.6

15. -2.5

EXERCISE 1-6

Solve as indicated. (Round answers to one decimal place, as needed.)

1. Find the length of the arc intercepted by a central angle of $\dfrac{\pi}{3}$ in a circle of radius 18 feet.

2. Find the length of the arc intercepted by a central angle of $225°$ in a circle of radius 12 meters.

3. Find the area of a sector intercepted by a central angle of $\frac{\pi}{3}$ in a circle of radius 18 feet.

4. Find the area of a sector intercepted by a central angle of 225° in a circle of radius 12 meters.

5. A wheel is turning at the rate of 960 revolutions per minute. Calculate its angular speed in radians per second.

2 Concepts from Geometry

MUST ⚡ KNOW

⚡ Knowing and applying the geometric properties related to triangles enables us to determine missing measures of angles and sides of triangles.

⚡ The triangle inequality theorem allows us to determine whether a triangle is possible when given three lengths for the sides.

⚡ The Pythagorean theorem allows us to find the length of the third side of a right triangle when we know the lengths of two of its sides, and we can use it to determine whether a triangle is a right triangle.

Before launching into trigonometry, there are a few ideas from geometry that you need to know well—namely, the properties of the angles and sides of triangles, the Pythagorean theorem, and similar triangles. These concepts are used repeatedly in trigonometry.

Furthermore, let's not forget that geometry is a rich topic on its own with countless applications in numerous areas in everyday life. It is widely used on a daily basis in fields from carpentry to chemistry, astronomy, and physics to robotics and computer gaming. It is used by architects to design buildings and by artists to design works of art.

IRL In an interview with the stained glass artist Kathryn Voigtel Welch, owner of River City Glassworks in San Marcos, Texas, the authors learned that much of the beauty in stained glass art owes itself to the mathematical principles of geometry. Looking at geometric examples of her exquisite work, we could see an array of various shapes, such as triangles, circles, semicircles, sectors, and quadrilaterals, all delicately put together to form beautiful patterns with perfect symmetry.

The Sum of a Triangle's Angles

The sum of the interior angles of a triangle is $180°$.

EXAMPLE

▶ In triangle ABC shown, solve for B.

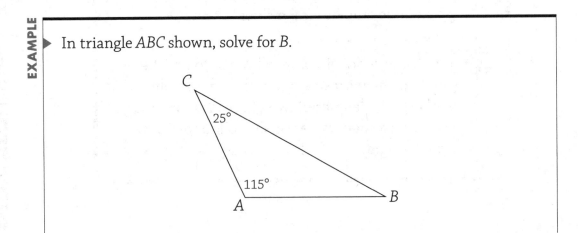

$$A + B + C = 180°$$
$$115° + B + 25° = 180°$$
$$B = 180° - 140°$$
$$B = 40°$$

The Triangle Inequality Theorem

The **triangle inequality theorem** asserts that the sum of the lengths of any two sides of a triangle is greater than the length of the third side.

EXAMPLE

▶ Can the lengths 5, 8, and 11 be the lengths of the sides of a triangle? Yes or No?

▶ The longest side is 11. Compare $5 + 8$ and 11.

$$5 + 8 > 11 \text{ True or False?}$$
$$13 > 11 \text{ True.}$$

▶ Because $13 > 11$, the answer is Yes—5, 8, and 11 can be the lengths of the sides of a triangle.

In simple terms, the triangle inequality theorem means that the length of any side of a triangle must be shorter than the sum of the lengths of the other two sides; otherwise, the three lengths will not make a triangle.

EXAMPLE

▶ Can the lengths 2, 6, and 9 be the lengths of the sides of a triangle? Yes or No?

▶ The longest side is 9. Compare $2 + 6$ and 9.

$$2 + 6 > 9 \text{ True or False?}$$
$$8 > 9 \text{ False.}$$

▶ Because $8 \not> 9$, the answer is No—2, 6, and 9 cannot be the lengths of the sides of a triangle.

The Pythagorean Theorem

In a right triangle, the **hypotenuse** is the side opposite the right angle and the other two sides are the **legs** (illustrated next).

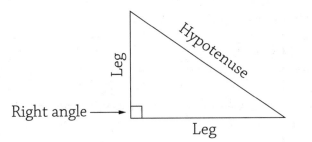

BTW

The hypotenuse is always the longest side of a right triangle.

- **Pythagorean theorem** In a right triangle, $c^2 = a^2 + b^2$, where c is the length of the hypotenuse and a and b are the lengths of the legs of the right triangle.

IRL Pythagoras is often credited with the discovery of the Pythagorean theorem, although, in truth, it was widely known before him. He also is credited with the theorem that the sum of the angles of a triangle is equal to two right angles. It is believed that around 525 BCE he founded the Pythagoreans, a religious brotherhood devoted to the study of numbers. This group knew the generalization which states that a polygon with n sides has a sum of interior angles equal to $2n - 4$ right angles and a sum of exterior angles equal to four right angles.

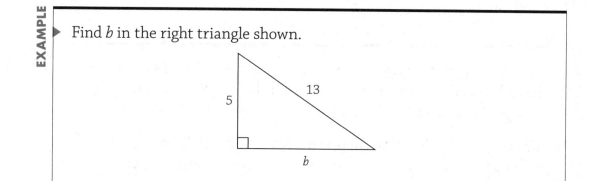

When we know the lengths of the two legs of a right triangle, we can use the Pythagorean theorem to find the length of the right triangle's hypotenuse.

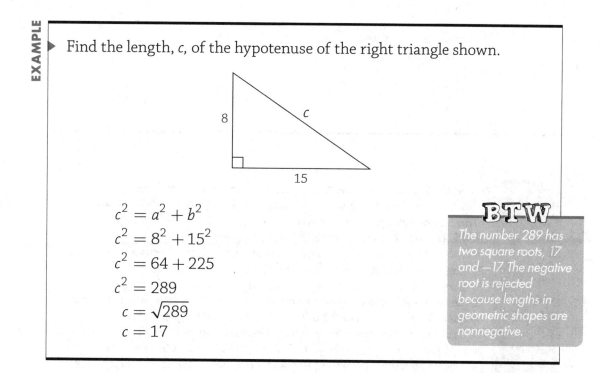

EXAMPLE

▶ Find the length, c, of the hypotenuse of the right triangle shown.

$$c^2 = a^2 + b^2$$
$$c^2 = 8^2 + 15^2$$
$$c^2 = 64 + 225$$
$$c^2 = 289$$
$$c = \sqrt{289}$$
$$c = 17$$

BTW

The number 289 has two square roots, 17 and −17. The negative root is rejected because lengths in geometric shapes are nonnegative.

If we know the lengths of the hypotenuse and one leg of a right triangle, we can use the Pythagorean theorem to find the length of the other leg.

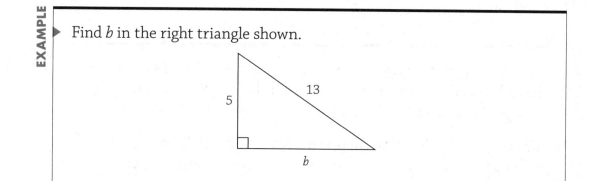

EXAMPLE

▶ Find b in the right triangle shown.

$$c^2 = a^2 + b^2$$
$$13^2 = 5^2 + b^2$$
$$169 = 25 + b^2$$
$$169 - 25 = 25 + b^2 - 25$$
$$144 = b^2$$
$$\sqrt{144} = b$$
$$12 = b$$

IRL A **Pythagorean triple** is three positive integers a, b, and c for which $a^2 + b^2 = c^2$. Commonly, we write such a triple as (a, b, c). Well-known examples of Pythagorean triples are (3, 4, 5), (5, 12, 13), (7, 24, 25), and (8, 15, 17). What's more, if (a, b, c) is a Pythagorean triple, then so is (na, nb, nc) for any positive integer n. As you have likely surmised, the name Pythagorean triple is derived from the Pythagorean theorem.

Do you know we even have days that are celebrated as Pythagorean Theorem Days? These days occur on a date in which the square of the month number plus the square of the day number equals the square of the last two digits of the year number. So, you might imagine that they don't occur every year. Three such days are August 15, 2017 (8/15/17) because $8^2 + 15^2 = 17^2$; December 16, 2020 (12/16/20) because $12^2 + 16^2 = 20^2$; and July 24, 2025 (7/24/25) because $7^2 + 24^2 = 25^2$.

In everyday life, we often encounter instances in which the Pythagorean theorem can be applied.

EXAMPLE

▶ Two people want to carry a tall mirror that is a 9 by 9 feet square through a doorway that measures 3 feet by 8 feet. Will the mirror fit through the doorway?

▶ The doorway is only 8 feet tall, so the two people will have to tilt the mirror and try to slide it through the doorway at an angle. It will fit through the doorway only if the doorway's diagonal is at least 9 feet in length. So, let's calculate the length of the doorway's diagonal.

$$c^2 = a^2 + b^2$$
$$c^2 = 3^2 + 8^2$$
$$c^2 = 9 + 64$$
$$c^2 = 73$$
$$c = \sqrt{73}$$
$$c \approx 8.5 \text{ feet}$$

▶ Therefore, because 8.5 feet is less than 9 feet, the mirror will not fit through the doorway.

 IRL The Pythagorean theorem is used extensively in the real world. It is used in architecture and construction, in navigation, and in surveying, just to name a few fields.

The Pythagorean theorem's converse is also true: In a triangle, if $c^2 = a^2 + b^2$, where c is the length of the triangle's longest side and a and b are the lengths of the triangle's other two sides, the triangle is a right triangle, and the right angle is opposite the longest side.

EXAMPLE

▶ Is a triangle with sides of lengths 3, 4, and 5 a right triangle? Yes or No?

▶ The longest side is 5. Let's compare 5^2 and $3^2 + 4^2$.

$$5^2 = 3^2 + 4^2 \text{ True or False?}$$
$$25 = 9 + 16 \text{ True or False?}$$
$$25 = 25 \text{ True.}$$

▶ Because $5^2 = 3^2 + 4^2$, the answer is Yes—a triangle with sides of lengths 3, 4, and 5 is a right triangle.

Similar Triangles

As illustrated next, two triangles are similar (denoted $\sim$) if their corresponding angles have the same measure and their corresponding sides are proportional in length.

Note: It is customary to write the corresponding vertices in the same order.

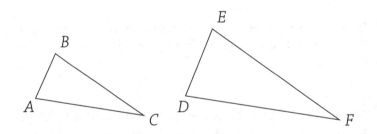

Given $\triangle ABC \sim \triangle DEF$, then (1) $\angle A$ and $\angle D$ have the same measure, $\angle B$ and $\angle E$ have the same measure, and $\angle C$ and $\angle F$ have the same measure, and (2) $\dfrac{AB}{DE} = \dfrac{BC}{EF} = \dfrac{AC}{DF}$.

Either of these two conditions guarantees that two triangles are similar. As a result, two triangles are similar if either two angles of one are equal in measure to two angles of the other or if the ratios of the lengths of their corresponding sides are equal.

When we are given two similar triangles, we can use the relationships that exist between the corresponding angles and sides to find missing angle measures or side lengths.

EXAMPLE

▶ In the following figure, $\triangle ABC \sim \triangle PQR$, find the measure of $\angle R$ and the length x.

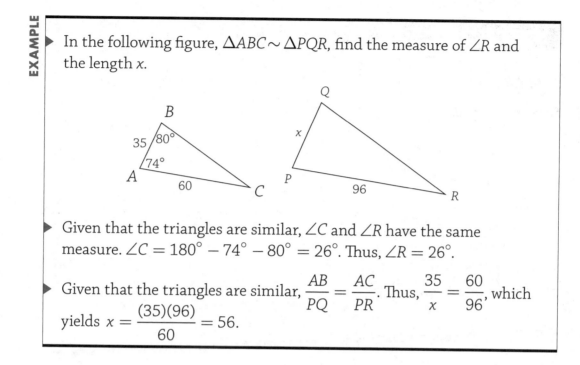

▶ Given that the triangles are similar, $\angle C$ and $\angle R$ have the same measure. $\angle C = 180° - 74° - 80° = 26°$. Thus, $\angle R = 26°$.

▶ Given that the triangles are similar, $\dfrac{AB}{PQ} = \dfrac{AC}{PR}$. Thus, $\dfrac{35}{x} = \dfrac{60}{96}$, which yields $x = \dfrac{(35)(96)}{60} = 56$.

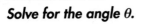

EXERCISES

EXERCISE 2-1

Solve for the angle θ.

1.

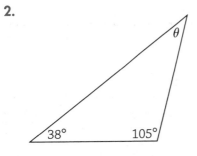

2.

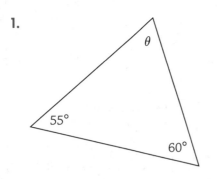

3.

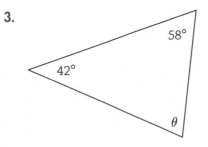

4.

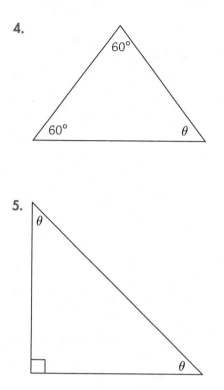

5.

EXERCISE 2-2

State Yes or No as to whether the given lengths can be the lengths of the sides of a triangle. Justify your answer.

1. 8, 16, 22

2. 3, 4, 5

3. 8, 11, 2

4. 1, 1, 1

5. 502, 21, 485

EXERCISE 2-3

For questions 1 to 4, find the missing side length in the right triangle.

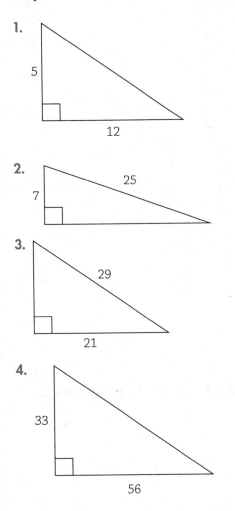

1.

5

12

2.

25

7

3.

29

21

4.

33

56

For questions 5 to 7, state Yes or No as to whether the set of numbers could be the lengths of the sides of a right triangle. Justify your answer.

5. 13, 84, 85

6. 39, 80, 89

7. 17, 55, 41

For questions 8 to 10, solve as indicated. (Round answers to one decimal place, as needed.)

8. The lower end of a 25-foot pole that is leaning against a wall is 7 feet from the base of a building. At what height is the top of the pole touching the building?

9. An 18-foot lamp pole casts a shadow of 28 feet. What is the distance from the top of the pole to the tip of the shadow?

10. Find the length of a diagonal of a rectangular garden that has dimensions of 30 feet by 40 feet.

EXERCISE 2-4

For questions 1 to 3, state Yes or No as to whether the two triangles are similar. Justify your answer.

1.

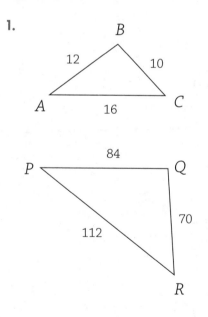

2.

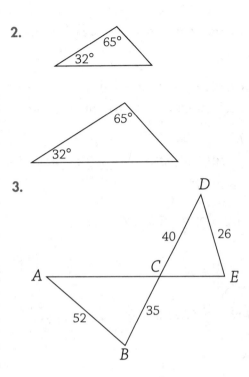

3.

For questions 4 and 5, solve as indicated.

4. In the following figure, $\triangle ABC \sim \triangle TUV$, find x.

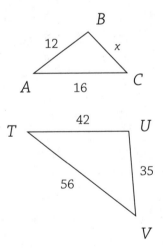

5. Find x.

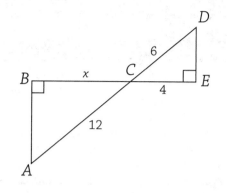

Flashcard App

Right Triangle Trigonometry

MUST ⚡ KNOW

⚡ Right triangle trigonometry allows us to see the relationships between the angles and sides of right triangles.

⚡ The most frequently used trigonometric ratios are the sine, cosine, and tangent.

⚡ When we are determining the trigonometric ratios of an acute angle θ, we can use any right triangle that contains θ as one of its angles.

⚡ Two very important right triangles are the $30° - 60° - 90°$ right triangle and the $45° - 45° - 90°$ right triangle.

ur study of trigonometry begins with looking at the relationships between the sides and angles of triangles. As we continue on this journey through the intriguing domain of trigonometry, you will find that the trigonometric ratios associated with an acute angle are basic concepts that you will encounter again and again. Knowing the trigonometric ratios of the sides of a right triangle associated with an acute angle of the right triangle is key to the study of trigonometry.

IRL The word *trigonometry* is derived from two Greek words: *trigono*, meaning "triangle," and *metro* meaning "measure." It is well known that the ancient Greeks, in particular Hipparchus and Ptolemy, used trigonometry in their study of astronomy.

Trigonometric Ratios of an Acute Angle in a Right Triangle

Recalling from Chapter 1, we know that the sum of the angles of a triangle is 180°. In a right triangle, one of the angles is 90° and the other two angles are acute angles whose sum is 90° (that is, they are complementary angles).

Consider a right triangle ABC, with the right angle at C and sides of lengths a, b, and c. The triangle's **hypotenuse** is $\overline{AB}$, the side opposite the right angle, and has length c. Relative to the acute angle A, the leg $\overline{BC}$ is its **opposite** side with length a, and the leg $\overline{AC}$ is its **adjacent** side with length b (illustrated next).

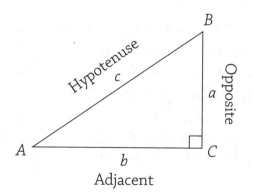

Note: When no confusion is likely to occur, you can refer to the angles of a triangle by their vertex points. For example, in the right triangle shown earlier, let's refer to the angle *BAC* simply as angle *A*.

 **IRL** Archaeologists using ground-penetrating radar mark the location of a subsurface object displaying a right angle because they are pretty certain they have located something built by humans. An archaeologist friend explained to the authors that right-angle archaeological objects are seldom naturally occurring, so their presence indicates a possibility of human involvement.

The six **trigonometric ratios** of angle *A* are defined as shown in the following box:

Name of Ratio	Abbreviation	Definition
sine *A*	sin *A*	$\sin A = \dfrac{\text{opposite}}{\text{hypotenuse}} = \dfrac{a}{c}$
cosine *A*	cos *A*	$\cos A = \dfrac{\text{adjacent}}{\text{hypotenuse}} = \dfrac{b}{c}$
tangent *A*	tan *A*	$\tan A = \dfrac{\text{opposite}}{\text{adjacent}} = \dfrac{a}{b}$
cosecant *A*	csc *A*	$\csc A = \dfrac{\text{hypotenuse}}{\text{opposite}} = \dfrac{c}{a}$
secant *A*	sec *A*	$\sec A = \dfrac{\text{hypotenuse}}{\text{adjacent}} = \dfrac{c}{b}$
cotangent *A*	cot *A*	$\cot A = \dfrac{\text{adjacent}}{\text{opposite}} = \dfrac{b}{a}$

Ordinarily, we use the abbreviated names of the ratios. Also, instead of using the angle notation ∠*A* to denote an angle, we often use just a capital letter by itself, e.g., *A*, *B*, or *C*, a lowercase variable name, e.g., *x*, *y*, or *t*, or letters from the Greek alphabet, e.g., α (alpha), β (beta), γ (gamma), θ (theta), or ϕ (phi), to represent angles.

BTW

*A mnemonic for remembering the sine, cosine, and tangent ratios is "soh-cah-toa" (soh-kuh-toh-uh), formed from the first letters of "**s**ine is **o**pposite over **h**ypotenuse; **c**osine is **a**djacent over **h**ypotenuse; and **t**angent is **o**pposite over **a**djacent."*

EXAMPLE

▶ Find the six trigonometric ratios for angle A in the right triangle ABC shown next.

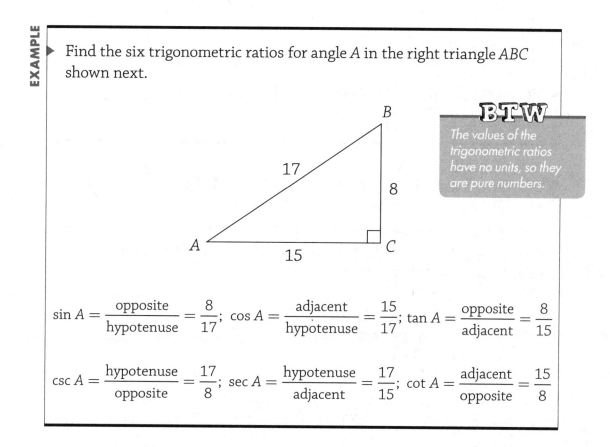

BTW

The values of the trigonometric ratios have no units, so they are pure numbers.

$$\sin A = \frac{\text{opposite}}{\text{hypotenuse}} = \frac{8}{17}; \quad \cos A = \frac{\text{adjacent}}{\text{hypotenuse}} = \frac{15}{17}; \quad \tan A = \frac{\text{opposite}}{\text{adjacent}} = \frac{8}{15}$$

$$\csc A = \frac{\text{hypotenuse}}{\text{opposite}} = \frac{17}{8}; \quad \sec A = \frac{\text{hypotenuse}}{\text{adjacent}} = \frac{17}{15}; \quad \cot A = \frac{\text{adjacent}}{\text{opposite}} = \frac{15}{8}$$

You might have noticed that the pairs $\sin A$ and $\csc A$, $\cos A$ and $\sec A$, and $\tan A$ and $\cot A$ are reciprocals. It turns out that for any given angle θ we have the following **reciprocal relationships**:

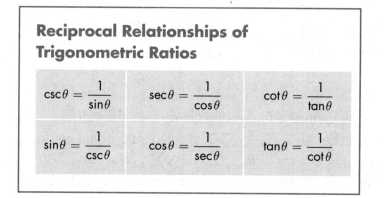

Reciprocal Relationships of Trigonometric Ratios

$\csc\theta = \dfrac{1}{\sin\theta}$	$\sec\theta = \dfrac{1}{\cos\theta}$	$\cot\theta = \dfrac{1}{\tan\theta}$
$\sin\theta = \dfrac{1}{\csc\theta}$	$\cos\theta = \dfrac{1}{\sec\theta}$	$\tan\theta = \dfrac{1}{\cot\theta}$

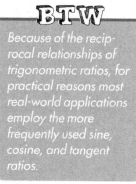

BTW

Because of the reciprocal relationships of trigonometric ratios, for practical reasons most real-world applications employ the more frequently used sine, cosine, and tangent ratios.

▶ Given $\sin\theta = \dfrac{3}{5}$, find $\csc\theta$.

▶ Because sine and cosecant are reciprocals, $\csc\theta = \dfrac{1}{\sin\theta} = \dfrac{1}{{}^3\!/_5} = \dfrac{5}{3}$.

Trigonometric Ratios of Special Acute Angles

Two very important right triangles are the $30° - 60° - 90°$ right triangle and the $45° - 45° - 90°$ right triangle. These two right triangles contain special acute angles (illustrated next).

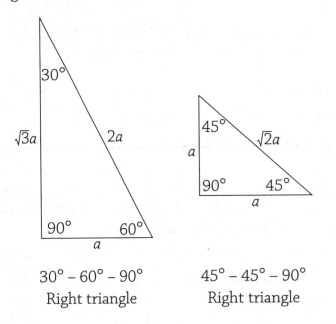

$30° - 60° - 90°$
Right triangle

$45° - 45° - 90°$
Right triangle

Note: a is a nonzero factor.

You might ask why did we include a factor of a for each of the sides of the triangles? In the case of the $30° - 60° - 90°$ right triangle, you could consider the triangle when $a = 1$ with sides 1, 2, and $\sqrt{3}$ as the "basic" $30° - 60° - 90°$ right triangle. Any triangle whose sides have a common factor times 1, 2, and $\sqrt{3}$ is similar to the basic triangle, and thus is a $30° - 60°$ $- 90°$ right triangle. Similarly, for the $45° - 45° - 90°$ right triangle, you could consider the triangle when $a = 1$ with sides 1, 1, and $\sqrt{2}$ as the "basic" $45° - 45° - 90°$ right triangle. Any triangle whose sides have a common factor times 1, 1, and $\sqrt{2}$ is similar to the basic triangle, and thus is a $45° - 45° - 90°$ right triangle.

In other words, right triangles whose sides have the same ratio as the sides (short leg:long leg:hypotenuse) of a $30° - 60° - 90°$ are $30° - 60° - 90°$ right triangles. Similarly, right triangles whose sides have the same ratio as the sides (leg:leg:hypotenuse) of a $45° - 45° - 90°$ are $45° - 45° - 90°$ right triangles.

Using the information shown in the figures, we can determine the following trigonometric ratios of the special acute angles 30°, 45°, and 60° $\left(\text{or } \dfrac{\pi}{6}, \dfrac{\pi}{4}, \text{ and } \dfrac{\pi}{3}, \text{ respectively} \right)$ associated with these two triangles.

BTW

Because from here on out we will frequently encounter these special acute angles, you should learn to construct the right triangles that contain them.

Sines, Cosines, and Tangents of Special Acute Angles

$\sin 30° = \sin\dfrac{\pi}{6} = \dfrac{1}{2}$	$\cos 30° = \cos\dfrac{\pi}{6} = \dfrac{\sqrt{3}}{2}$	$\tan 30° = \tan\dfrac{\pi}{6} = \dfrac{1}{\sqrt{3}}$ or $\dfrac{\sqrt{3}}{3}$
$\sin 45° = \sin\dfrac{\pi}{4} = \dfrac{1}{\sqrt{2}}$ or $\dfrac{\sqrt{2}}{2}$	$\cos 45° = \cos\dfrac{\pi}{4} = \dfrac{1}{\sqrt{2}}$ or $\dfrac{\sqrt{2}}{2}$	$\tan 45° = \tan\dfrac{\pi}{4} = 1$
$\sin 60° = \sin\dfrac{\pi}{3} = \dfrac{\sqrt{3}}{2}$	$\cos 60° = \cos\dfrac{\pi}{3} = \dfrac{1}{2}$	$\tan 60° = \tan\dfrac{\pi}{3} = \sqrt{3}$

All right triangles similar to the two special right triangles have the same trigonometric ratios as shown earlier. In fact, you can use the concept of similar triangles to establish that the trigonometric ratios associated with a given angle of a right triangle are constant, regardless of the lengths of the sides that form the angle. Therefore, when we are determining the trigonometric ratios of any acute angle θ, we can use any right triangle that contains θ as one of its angles.

EXAMPLE

▶ In triangle ABC with $C = 90°$, $c = 16$, and $a = 8$, what is the measure of angle A?

▶ Because $c = 16 = 2 \cdot 8 = 2a$, we can use the Pythagorean theorem

to determine that $b = \sqrt{(2a)^2 - a^2} = \sqrt{4a^2 - a^2} = \sqrt{3a^2} = \sqrt{3}a$. Therefore, the ratio of the sides (short leg:long leg:hypotenuse) is $a : \sqrt{3}a : 2a$. Thus, triangle ABC is a $30° - 60° - 90°$ right triangle in which a is opposite the $30°$ angle. Hence, the measure of angle A is $30°$.

Finding Missing Sides in Right Triangles with Special Acute Angles

In a given right triangle, if we are given the measure of one acute angle and the length of one side, we can use trigonometric ratios to determine missing side lengths in the triangle. For each missing side, we select the trigonometric ratio that has the unknown side length as either the numerator or the denominator. Next, we substitute the known and unknown information into the definition of the trigonometric ratio. Then we solve for the missing side length.

EXAMPLE

▶ In triangle ABC with $C = 90°$, find a, given $A = 30°$ and $c = 48$.

▶ We are given an acute angle and the length of the hypotenuse. The unknown, a, is opposite the angle, so we use the sine to determine a.

▶ $\sin 30° = \dfrac{a}{48}$; $\dfrac{1}{2} = \dfrac{a}{48}$, which yields $a = 24$.

In some cases, to give an exact answer, we might need to express the answer using a radical form.

▶ In triangle RST with $S = 90°$, find r, given $R = 60°$ and $t = 5$. Write the exact answer in the simplest radical form.

▶ We are given an acute angle and the adjacent side. The unknown, r, is the opposite side, so we use the tangent to determine r.

▶ $\tan 60° = \dfrac{r}{5}; \sqrt{3} = \dfrac{r}{5}$, which yields $r = 5\sqrt{3}$.

BTW

Being asked to find an "exact" answer when a problem involves evaluating the trigonometric function of an angle is usually a harbinger that special angles are in play. So, don't expect to use the trigonometric function keys on your calculator but, instead, get ready to use your knowledge of the trigonometric ratios associated with $30°$, $45°$, and $60°$ angles.

EXERCISES

EXERCISE 3-1

For questions 1 to 5, find the indicated trigonometric ratio for angle B using the following right triangle.

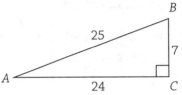

1. $\sin B$

2. $\sec B$

3. $\csc B$

4. $\cos B$

5. $\tan B$

For questions 6 to 10, answer as indicated.

6. Given $\csc\theta = \dfrac{41}{9}$, find $\sin\theta$.

7. Given $\tan\beta = \dfrac{12}{35}$, find $\cot\beta$.

8. Given $\sec\gamma = \dfrac{61}{11}$, find $\cos\gamma$.

9. Use the right triangle shown to determine $\tan\theta$.

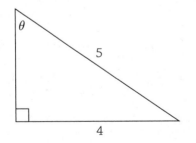

10. Use the right triangle shown to determine $\sec\beta$.

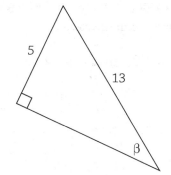

EXERCISE 3-2

Answer as indicated.

1. What is the simplest ratio of the sides (short leg:long leg:hypotenuse) in a $30° - 60° - 90°$ right triangle?

2. What is the simplest ratio of the sides (leg:leg:hypotenuse) in a $45° - 45° - 90°$ triangle?

3. In triangle ABC with $C = 90°$, $c = 12$, and $b = 6$, what is the measure of angle A?

4. In triangle ABC with $C = 90°$, $c = 15\sqrt{2}$, and $a = 15$, what is the measure of angle A?

5. In triangle ABC with $C = 90°$, $c = 14$, and $b = 7\sqrt{2}$, what is the measure of angle B?

EXERCISE 3-3

Solve for the exact missing side length in the right triangle. (Write the exact answer in the simplest radical form for irrational answers.)

1. In triangle ABC with $C = 90°$, find b, given $A = 30°$ and $c = 26$.

2. In triangle ABC with $C = 90°$, find a, given $A = 60°$ and $c = 48$.

3. In triangle ABC with $C = 90°$, find c, given $B = 30°$ and $b = 20$.

4. In triangle ABC with $C = 90°$, find a, given $A = 45°$ and $c = 10$.

5. In triangle ABC with $C = 90°$, find b, given $A = 60°$ and $a = 9$.

Flashcard App

 # The Trigonometry of General Right Triangles

MUST KNOW

⚡ Solving a triangle means determining all of its angle measures and the lengths of all of its sides.

⚡ The sine, cosine, and tangent ratios are used to solve for the measures of the sides and angles of a right triangle.

⚡ For problems involving angles other than special angles, we use calculators to find the values of trigonometric ratios.

⚡ The angle of elevation is the angle from the horizontal upward to a point of interest, while the angle of depression is the angle from the horizontal downward to a point of interest.

⚡ The trigonometry of general right triangles is often used as a means of indirectly determining distances or lengths by using measurements of angles and known distances.

The skill of solving right triangles is basic to mastery of trigonometry and its applications. There are tasks in the real world, such as building a wheelchair ramp, that can be modeled with a right triangle in which certain information about the sides and angles are known, and you need to find unknown measures of sides or angles. In such situations, understanding right triangle trigonometry can be indispensable.

This chapter (as in the previous chapter) involves solving for missing elements in right triangles. The difference in this chapter is that we will encounter angles other than the special angles introduced in Chapter 3. Therefore, we will find a calculator to be a valuable tool when we are working through the examples and exercises in this chapter. (See Appendix A for guidance in using the trigonometric features of the TI-84 Plus graphing calculator.)

Solving Right Triangles

A triangle has three sides and three angles. **Solving a triangle** means determining all of its angle measures and the lengths of all of its sides. In addition to knowing that a right triangle has one angle of 90°, we must either know the lengths of two sides or the length of one side and the measure of one acute angle in order to solve the triangle.

BTW

When solving triangles in this chapter, make sure your calculator is set to degree mode (see Appendix A for instructions).

EXAMPLE

▶ Solve the triangle shown next. Round answers to one decimal place, as needed.

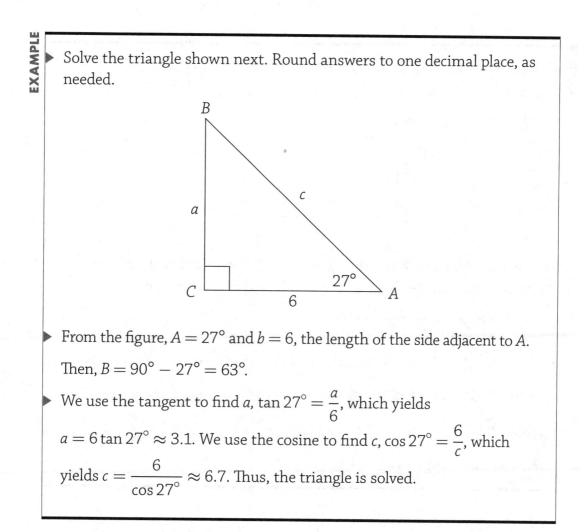

▶ From the figure, $A = 27°$ and $b = 6$, the length of the side adjacent to A. Then, $B = 90° - 27° = 63°$.

▶ We use the tangent to find a, $\tan 27° = \dfrac{a}{6}$, which yields $a = 6 \tan 27° \approx 3.1$. We use the cosine to find c, $\cos 27° = \dfrac{6}{c}$, which yields $c = \dfrac{6}{\cos 27°} \approx 6.7$. Thus, the triangle is solved.

When two sides of a right triangle are given, we use the Pythagorean theorem to solve for the third side. Then we use the relationships between the sides and angles of the triangle to determine the measures of the acute angles.

EXAMPLE

Solve the triangle shown next. Round answers to one decimal place, as needed.

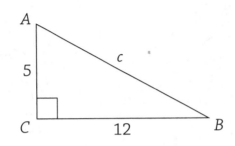

From the figure, we have $a = 12$ and $b = 5$, the length of the two legs of the right triangle. Then, $c^2 = 12^2 + 5^2 = 144 + 25 = 169$, from which we have $c = \sqrt{169} = 13$.

For A, $\tan A = \dfrac{12}{5} = 2.4$. (As shown in Appendix A, we determine A by using the inverse tangent calculator key sequence: $\boxed{2ND}[\tan^{-1}]$.)

Thus, $A = \tan^{-1}(2.4) \approx 67.4°$. Thereafter, $B \approx 90° - 67.4° = 22.6°$.

Hence, the triangle is solved.

Applications of Right Triangle Trigonometry

This lesson presents some of the ways trigonometry can be applied in the real world. Among the concepts we encounter are the angle of elevation and the angle of depression. **Angle of elevation** refers to the angle from the horizontal upward to a point of interest, while **angle of depression** refers to the downward angle formed between the horizontal and the line of sight to a point of interest below the horizontal (see the following illustration).

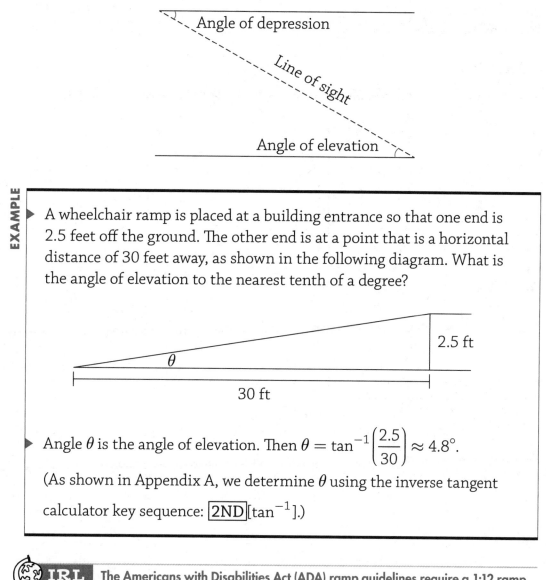

EXAMPLE

A wheelchair ramp is placed at a building entrance so that one end is 2.5 feet off the ground. The other end is at a point that is a horizontal distance of 30 feet away, as shown in the following diagram. What is the angle of elevation to the nearest tenth of a degree?

Angle θ is the angle of elevation. Then $\theta = \tan^{-1}\left(\dfrac{2.5}{30}\right) \approx 4.8°$.

(As shown in Appendix A, we determine θ using the inverse tangent calculator key sequence: $\boxed{\text{2ND}}[\tan^{-1}]$.)

IRL   The Americans with Disabilities Act (ADA) ramp guidelines require a 1:12 ramp slope ratio, which equals a slope of approximately 4.8 degrees, or 1 foot of horizontal wheelchair ramp distance for each 1 inch of rise.

EXERCISES

EXERCISE 4-1

Solve the given right triangle. (Round answers to one decimal place, as needed.)

1.

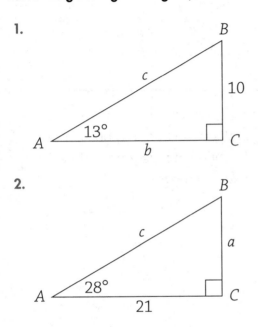

2.

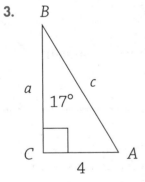

3.

4.

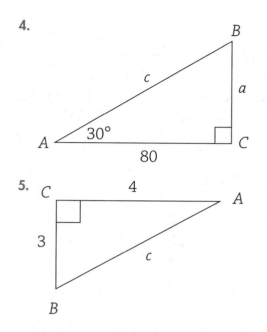

5.

EXERCISE 4-2

Solve as indicated. (Round answers to one decimal place, as needed.)

1. A 40-foot support wire on a TV dish satellite makes an angle of 60° with the ground. Calculate the vertical distance above the ground of the point where the wire touches the dish.

2. A high-dive platform 80 feet off the ground is supported by a guy wire 140 feet long that is attached to the ground to support the platform. Calculate the measure of the angle where the wire touches the ground.

3. A forester wants to estimate the height of a giant tree from a distance of 150 feet. The angle of elevation from the forester's position to the top of the tree is 52°. Determine the height of the tree.

4. A ham radio antenna stands on top of a house. The base of the antenna is 20 feet from the ground. The angle of depression from the top of the antenna to a point 100 feet from the base of the house measures 15°. Let x be the height of the antenna. Find x. (See the following diagram.)

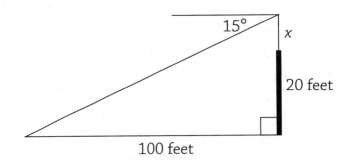

5. A kite is attached to a string that is 200 feet long. If the string makes an angle of 40° with the ground, how high is the kite?

Flashcard App

The Trigonometry of Oblique Triangles

MUST⚡KNOW

⚡ An oblique triangle has either three acute angles or one obtuse angle and two acute angles but no right angle.

⚡ The Law of Cosines applies when the problem provides the lengths of three sides or the lengths of two sides and the measure of their included angle.

⚡ The Law of Sines applies when the problem provides the measures of two sets of angles and the lengths of their opposite sides.

⚡ The ambiguous case arises when the only specifications given for a proposed triangle are the lengths of two sides and a non-included acute angle of the triangle.

aving a firm understanding of oblique triangle trigonometry is a
practical and necessary skill in trigonometry. An **oblique** triangle
is any triangle that does not contain a right angle. It could contain
three acute angles, in which case it is an **acute triangle**; or it could contain
an obtuse angle and two acute angles, in which case it is an **obtuse triangle**.
Oblique triangle trigonometry is somewhat more challenging than right
triangle trigonometry, although the basic concepts stay the same. Being able
to solve oblique triangles and find their areas is critical to the application
of trigonometry in engineering, architecture, navigation, surveying,
oceanography, astronomy, physics, and higher mathematics. To that end, we
rely mainly on two useful laws of trigonometry, namely, the Law of Cosines
and the Law of Sines.

IRL In a physical sense, *oblique* is used to describe something that is slanting or
inclined. Figuratively, *oblique* means "indirect" or "purposely misleading,"
such as when a person gives an evasive answer to a question.

Law of Cosines (SAS or SSS)

The Law of Cosines consists of formulas for solving oblique triangles when
we are given the lengths of two sides and the included angle between the
two sides (abbreviated SAS) or the lengths of all three sides (abbreviated
SSS) of a triangle.

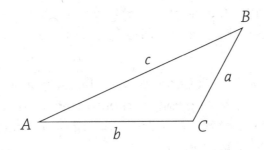

For the triangle shown in the previous illustration, the **Law of Cosines** is as follows:

$$a^2 = b^2 + c^2 - 2bc \cos A$$
$$b^2 = a^2 + c^2 - 2ac \cos B$$
$$c^2 = a^2 + b^2 - 2ab \cos C$$

BTW

The Law of Cosines also can be used when you have a right triangle. In fact, the Law of Cosines' formulas constitute a generalization of the Pythagorean theorem.

EXAMPLE

▶ Find c as shown in the following figure. Round answers to one decimal place, as needed.

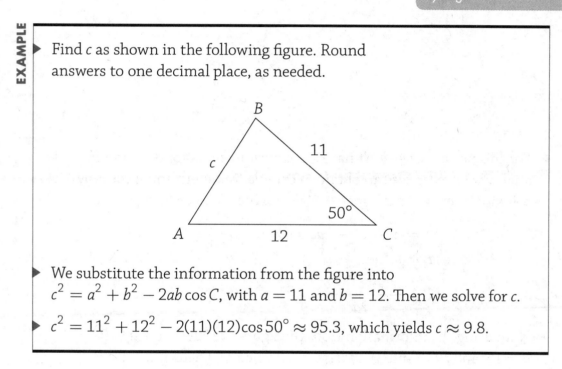

▶ We substitute the information from the figure into $c^2 = a^2 + b^2 - 2ab \cos C$, with $a = 11$ and $b = 12$. Then we solve for c.

▶ $c^2 = 11^2 + 12^2 - 2(11)(12)\cos 50° \approx 95.3$, which yields $c \approx 9.8$.

Let's try another example.

▶ Find B as shown in the following figure. Round the answer to one decimal place, as needed.

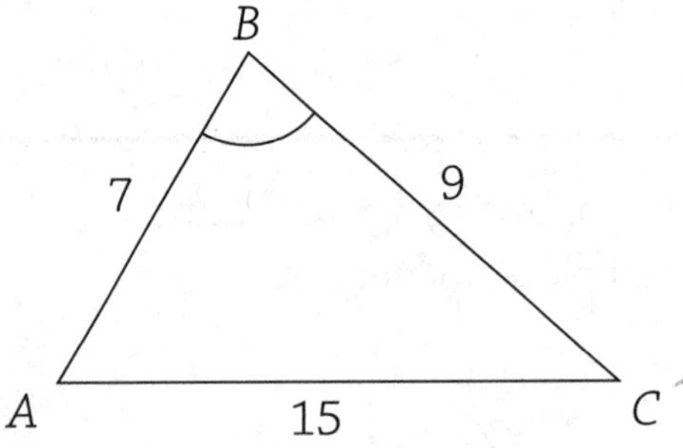

▶ For this problem, we don't have an explicit formula for determining angle B. However, we can obtain a formula for determining $\cos B$ by solving the equation $b^2 = a^2 + c^2 - 2ac \cos B$ for $\cos B$:

$$b^2 = a^2 + c^2 - 2ac \cos B$$
$$2ac \cos B = a^2 + c^2 - b^2$$
$$\cos B = \frac{a^2 + c^2 - b^2}{2ac}$$

▶ Now, using the information from the figure, we substitute $a = 9$, $b = 15$, and $c = 7$ into the formula obtained and determine $\cos B$.

$$\cos B = \frac{a^2 + c^2 - b^2}{2ac}$$

$$\cos B = \frac{9^2 + 7^2 - 15^2}{2(9)(7)}$$

$$\cos B \approx -0.7540$$

▶ Next, we use the $\boxed{\text{2ND}}$ [$\cos^{-1}$] calculator key sequence to determine B.

$$B \approx \cos^{-1}(-0.7540) \approx 138.9°.$$

Let's look at an example of an application of the Law of Cosines.

EXAMPLE

▶ A **resultant force** is the single force obtained when two or more forces act at a point concurrently. Two forces of 30 pounds and 45 pounds act on an object with an angle of 60° between them. Find the magnitude of the resultant force to the nearest pound.

▶ For a number of forces applied at the same point at the same time, we use arrows to represent the various forces, and we obtain the arrow that represents the resultant force by arranging the arrows representing the forces initial point to tip and then connecting the initial point of the first arrow to the tip of the last arrow. You might be surprised to learn that you get the same resultant regardless of the order of the arrangement. For two forces acting at a point concurrently, as in the current problem, the arrangement of the two forces forms a parallelogram.

▶ Let $\overrightarrow{AC}$ represent the 30-pound force and $\overrightarrow{CB}$ represent the 45-pound force. Sketch a parallelogram containing $\overrightarrow{AC}$ and $\overrightarrow{CB}$ placed initial point to tip. Draw $\overrightarrow{AB}$ as the resultant force that connects the initial point of $\overrightarrow{AC}$ to the tip of $\overrightarrow{CB}$ in a straight line, as shown next.

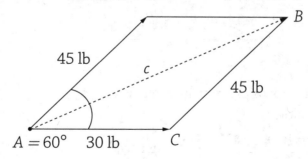

▶ Because A and C are consecutive angles of a parallelogram, they are supplementary. Thus, $C = 120°$. Now, omitting the units for convenience, we use the Law of Cosines, with $a = 30$ and $b = 45$, to find the magnitude, c, of the resultant force $\overrightarrow{AB}$.

$$c^2 = a^2 + b^2 - 2ab \cos C$$
$$c^2 = 30^2 + 45^2 - 2(30)(45) \cos(120°)$$
$$c^2 = 900 + 2{,}025 - 2{,}700 \cos(120°)$$
$$c^2 = 4{,}275$$
$$c = \sqrt{4{,}275} \approx 65.4$$

▶ Therefore, to the nearest pound, the magnitude of the resultant force is 65 pounds.

Law of Sines (ASA or AAS)

The **Law of Sines** consists of equations for solving oblique triangles when you are given two angles and the included side (abbreviated ASA) or two angles and a non-included side (abbreviated AAS) of a triangle.

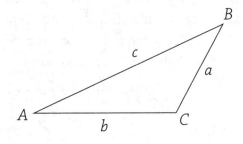

For the triangle shown previously, the Law of Sines can be stated as follows:

$$\frac{\sin A}{a} = \frac{\sin B}{b} = \frac{\sin C}{c} \quad \text{or, equivalently,}$$

$$\frac{a}{\sin A} = \frac{b}{\sin B} = \frac{c}{\sin C}$$

BTW

As with the Law of Cosines, the Law of Sines applies to right triangles as well as to oblique triangles.

We also can express the Law of Sines as the following set of three equations:

$$\frac{a}{b} = \frac{\sin A}{\sin B}, \frac{a}{c} = \frac{\sin A}{\sin C}, \text{ and } \frac{b}{c} = \frac{\sin B}{\sin C}$$

These three equations accentuate that the Law of Sines means that the sides of a triangle are proportional to the sines of their opposite angles.

EXAMPLE

▶ Given triangle ABC, with $B = 36°$, $C = 104°$, and $b = 12$, find c. Round the answer to one decimal place, as needed.

▶ Let's draw a sketch.

▶ We substitute the given information into $\dfrac{\sin B}{b} = \dfrac{\sin C}{c}$ and then solve for c.

$$\frac{\sin 36°}{12} = \frac{\sin 104°}{c}$$

$$c = \frac{12 \sin 104°}{\sin 36°} \approx 19.8$$

As you will see in the next example, sometimes a preliminary calculation is needed before proceeding with the Law of Sines.

EXAMPLE

▶ Given triangle ABC with $A = 40°$, $C = 104°$, and $b = 12$, solve for a. Round the answer to one decimal place, as needed.

▶ Let's draw a sketch.

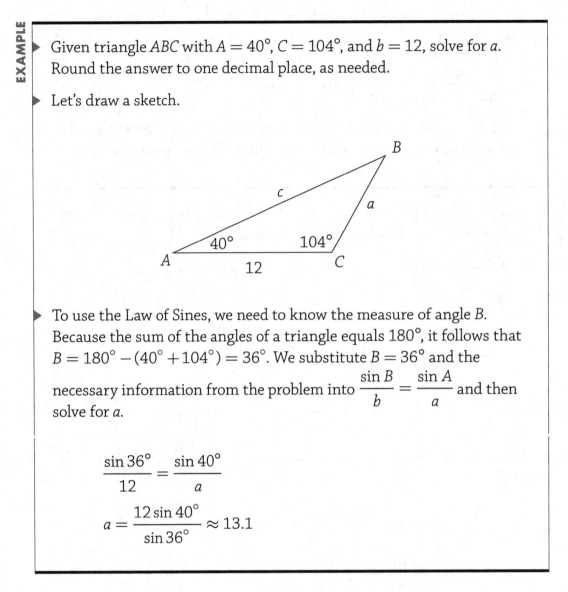

▶ To use the Law of Sines, we need to know the measure of angle B. Because the sum of the angles of a triangle equals $180°$, it follows that $B = 180° - (40° + 104°) = 36°$. We substitute $B = 36°$ and the necessary information from the problem into $\dfrac{\sin B}{b} = \dfrac{\sin A}{a}$ and then solve for a.

$$\frac{\sin 36°}{12} = \frac{\sin 40°}{a}$$

$$a = \frac{12 \sin 40°}{\sin 36°} \approx 13.1$$

In the next example, we use the Law of Sines to determine a missing measure of an angle.

EXAMPLE

In the following figure, find the measure of angle θ. Round the answer to one decimal place, as needed.

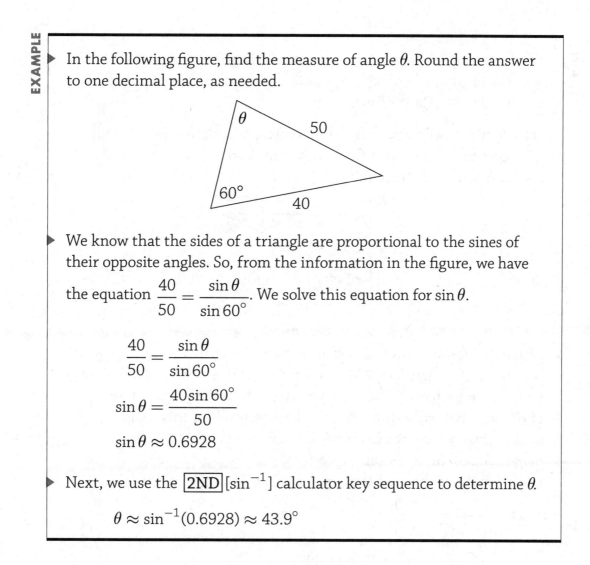

We know that the sides of a triangle are proportional to the sines of their opposite angles. So, from the information in the figure, we have the equation $\dfrac{40}{50} = \dfrac{\sin\theta}{\sin 60°}$. We solve this equation for $\sin\theta$.

$$\frac{40}{50} = \frac{\sin\theta}{\sin 60°}$$

$$\sin\theta = \frac{40\sin 60°}{50}$$

$$\sin\theta \approx 0.6928$$

Next, we use the $\boxed{\text{2ND}}\,[\sin^{-1}]$ calculator key sequence to determine θ.

$$\theta \approx \sin^{-1}(0.6928) \approx 43.9°$$

Now let's look at an example of an application of the Law of Sines involving a resultant.

▶ Two forces applied on a heavy carton make angles of $35°$ and $48°$ with the resultant force of 150 pounds. Find to the nearest pound the magnitude of the larger force.

▶ Let's draw a parallelogram sketch to represent the given forces, with $\overrightarrow{AC}$ representing the larger force with magnitude, V, and $\overrightarrow{AB}$ as the resultant force of 150 pounds.

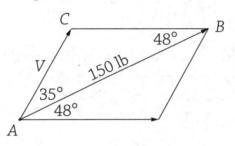

▶ Alternate interior angles of parallel lines are congruent, and the sum of the angles of a triangle is 180°. Hence, $C = 180° - (35° + 48°) = 97°$. Recall from geometry that in any triangle, the relative magnitude of the sides corresponds to the relative magnitude of their opposite angles. Thus, we use the Law of Sines to find V because it is the magnitude of the larger force (given that it is opposite the $48°$ angle, making it larger than the force opposite the $35°$ angle in triangle ABC).

Thus, $\dfrac{V}{\sin 48°} = \dfrac{150}{\sin 97°}$.

▶ Solving for V yields $V = \dfrac{150\sin(48°)}{\sin(97°)} \approx 112.3$.

▶ Hence, to the nearest pound, the magnitude of the larger force is 112 pounds.

Law of Sines Ambiguous Case (SSA)

When you are solving oblique triangles, the ambiguous case arises when the only specifications given for a proposed triangle are the lengths of two sides and a non-included acute angle of the triangle (abbreviated SSA). In this circumstance, three situations can occur: (1) no triangle exists, (2) one unique triangle exists, or (3) two distinct triangles may satisfy the conditions given.

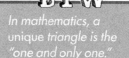

In mathematics, a unique triangle is the "one and only one."

Consider a proposed triangle in which you are given a, c, and A. If $A \geq 90°$ and $a > c$, then there exists one unique triangle; but if $a \leq c$, no such triangle exists.

If A is acute, then there are three possibilities: no triangle, one unique triangle, or two distinct triangles. Here is a visual depiction of the dilemma that occurs (where $h = c \sin A$ is the distance from angle B's vertex to the opposite side):

*Notice that the possibility of two triangles occurs **only** when you are given two sides and a non-included acute angle.*

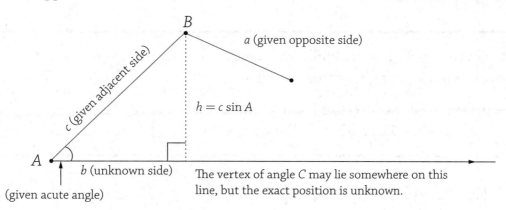

B

a (given opposite side)

c (given adjacent side)

$h = c \sin A$

A

b (unknown side)

(given acute angle)

The vertex of angle C may lie somewhere on this line, but the exact position is unknown.

As a result, the following outcomes are possible (illustrated on the next page):

- If $a \geq c$, a unique triangle exists.
- If $a = h$, one unique right triangle exists.
- If $a < h$, no such triangle exists.
- If $h < a < c$, two distinct triangles exist.

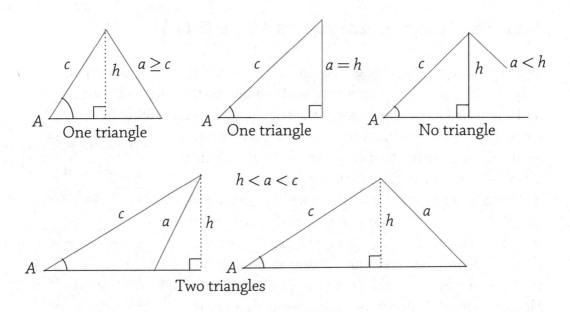

Let's look at three examples where you are given two sides and a non-included acute angle. Here is the first example.

EXAMPLE

▶ Given $A = 38°$, $c = 12$, and $a = 7$, decide whether there is no triangle, one unique triangle, or two triangles.

▶ A is acute, $a < c$ (because $7 < 12$), and $a < h = c \sin A$ (because $7 < 12 \sin 38° \approx 7.4$). Because $7 < h$, no triangle exists.

Now let's try the second example.

EXAMPLE

▶ Given $A = 42°$, $c = 12$, and $a = 22$, decide whether there is no triangle, one unique triangle, or two triangles.

▶ A is acute and $a > c$ (because $22 > 12$). Thus, there is one unique triangle.

Here's the third example.

EXAMPLE

▶ Given $A = 30°$, $c = 12$, and $a = 7$, decide whether there is no triangle, one unique triangle, or two triangles.

▶ A is acute, $a < c$ (because $7 < 12$), and $a > h = c \sin A$ (because $7 > 12 \sin 30° = 6$). Seeing as $h < 7 < c$, two triangles exist.

Now let's consider two examples where you are given two sides and a non-included obtuse angle. Here is the first example.

EXAMPLE

▶ Given $A = 100°$, $c = 16$, and $a = 10$, decide whether there is one unique triangle or no triangle.

▶ A is obtuse, $a < c$ (because $10 < 16$), so no triangle exists.

Let's try the next example.

EXAMPLE

▶ Given $A = 135°$, $c = 15$, and $a = 24$, decide whether there is one unique triangle or no triangle.

▶ A is obtuse and $a > c$ (because $24 > 15$), so there is one unique triangle.

Solving General Triangles

The Law of Cosines and the Law of Sines apply to both right triangles and oblique triangles. The following table is a guide to solving triangles. It presents a situation; the number of triangles that result; and whether the Law of Cosines, the Law of Sines, or neither is most appropriate for the situation.

BTW

Keep in mind that for a triangle to exist, the information given must not violate the triangle inequality theorem or the constraint that a triangle's interior angles must add up to 180°.

Information Given in the Problem	Number of Triangles	Use
Three sides (SSS) and the sum of the lengths of the two smaller sides is greater than the length of the larger side.	One unique triangle	Law of Cosines
Three sides (SSS) and the sum of the lengths of the two smaller sides is less than or equal to the length of the larger side.	No triangle	Neither
Two sides and the included angle (SAS).	One unique triangle	Law of Cosines
Two angles and the included side (ASA) and the sum of the given angles is less than 180°.	One unique triangle	Law of Sines
Two angles and a non-included side (AAS) and the sum of the given angles is less than 180°.	One unique triangle	Law of Sines
Two angles and either the included side (ASA) or a non-included side (AAS), and the sum of the given angles is greater than or equal to 180°.	No triangle	Neither
Two sides and a non-included obtuse angle (SSA) and the length of the side opposite the given angle is greater than the length of the side adjacent to the given angle.	One unique triangle	Law of Sines
Two sides and a non-included obtuse angle (SSA) and the length of the side opposite the given angle is less than or equal to the length of the side adjacent to the given angle.	No triangle	Neither

Information Given in the Problem	Number of Triangles	Use
Two sides and a non-included acute angle (SSA) and the length of the side opposite the given angle is greater than or equal to the length of the side adjacent to the given angle.	One unique triangle	Law of Sines
Two sides and a non-included acute angle (SSA) and the length of one side falls between the length of the altitude from the vertex where the two given sides meet and the length of the other side.	Two triangles	Law of Sines
Two sides and a non-included acute angle (SSA) and the length of the altitude from the vertex where the two given sides meet falls between the lengths of the two given sides.	No triangle	Neither
Three angles (AAA).	No unique triangle	Neither

Keep in mind that, when looking at the examples and working through the questions that follow, there can be multiple ways to find the solution in a given situation. You might think of ways to reach the correct solutions other than the ones shown.

EXAMPLE

▶ Solve triangle ABC, given $A = 40°$, $a = 50$, $b = 30$. Round answers to one decimal place, as needed.

▶ We are given two sides and a non-included acute angle (SSA), and the length of the side opposite the given angle is greater than the length of the side adjacent to the given angle, so there is one unique triangle.

▶ Using the Law of Sines yields $\dfrac{30}{\sin B} = \dfrac{50}{\sin 40°}$. Hence,

$\sin B = \dfrac{30 \sin 40°}{50} \approx 0.3857$. Given $\sin^{-1}(0.3857) \approx 22.7°$, then either $B \approx 22.7°$ or $B \approx 180° - 2.7° = 157.3°$. The latter value will not work because $40° + 157.3° > 180°$. It follows that $B \approx 22.7°$ and, thus, $C \approx 180° - 40° - 22.7° = 117.3°$.

▶ Using this result and again applying the Law of Sines gives

$\dfrac{50}{\sin 40°} = \dfrac{c}{\sin 117.3°}$. Hence, $c = \dfrac{50 \sin 117.3°}{\sin 40°} \approx 69.1$. Therefore, the triangle is solved.

Let's try another example.

EXAMPLE

▶ Solve triangle ABC, given $a = 15$, $b = 25$, and $c = 28$. Round answers to one decimal place, as needed.

▶ We are given three sides (SSS), and the sum of the lengths of the two smaller sides is greater than the length of the larger side, so there is one unique triangle. Applying the Law of Cosines, we solve for A.

$$\cos A = \frac{25^2 + 28^2 - 15^2}{2(25)(28)} \approx 0.8457$$

$$A \approx 32.3°$$

▶ Applying the Law of Cosines a second time, we solve for B.

$$\cos B = \frac{15^2 + 28^2 - 25^2}{2(15)(28)} \approx 0.4571$$

$$B \approx 62.8°$$

▶ Then $C \approx 180° - 32.3° - 62.8° = 84.9°$. Therefore, the triangle is solved.

There might be a trick in the following example. Be careful!

EXAMPLE

▶ Solve triangle ABC, given $A = 40°$, $B = 140°$, and $b = 30$.

▶ You are given two angles and the non-included side (AAS), and the sum of the given angles, $40°$ and $140°$, equals $180°$. So, there is no triangle.

IRL   There is also a Law of Tangents that relates the tangents of two angles of a triangle to the lengths of their opposite sides:

$$\frac{a - b}{a + b} = \frac{\tan\left(\dfrac{A - B}{2}\right)}{\tan\left(\dfrac{A + B}{2}\right)}$$

Before the proliferation of calculators and computers, the Law of Tangents was popular for its use with slide rules. Nowadays, along with the slide rule, the Law of Tangents is, for the most part, obsolete.

Area of a General Triangle Using Trigonometry

The familiar formula for the area of a triangle is Area $= \dfrac{1}{2}bh$, where b is the length of one side and h is the height of the triangle drawn to that side (or an extension of it). Using trigonometry, we can determine that for a general triangle ABC, with given sides a and b and the included angle C, $h = a \sin C$ (illustrated next). Substituting into the formula Area $= \dfrac{1}{2}bh$ yields Area $= \dfrac{1}{2}b$.

$a \sin C = \dfrac{1}{2}ab \sin C$, where a and b are any two sides of a triangle and C is the included angle.

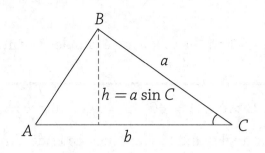

This formula works for all triangles as long as we have two sides and the included angle.

Let's apply the formula to an acute triangle.

▶ Find the area of the triangle shown next.

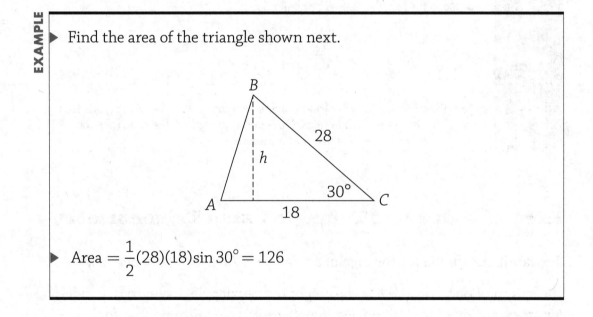

▶ Area $= \dfrac{1}{2}(28)(18)\sin 30° = 126$

Now, let's try the formula with an obtuse triangle.

EXAMPLE

▶ Find the area of the triangle shown next.

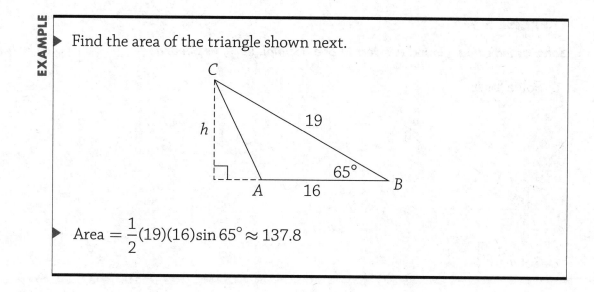

▶ Area $= \dfrac{1}{2}(19)(16)\sin 65° \approx 137.8$

EXERCISES

EXERCISE 5-1

Solve as indicated. (Round answers to one decimal place, as needed.)

1. Solve for b.

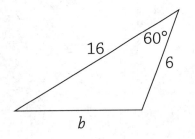

2. Solve for θ.

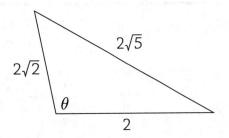

3. Solve triangle ABC given $B = 110°$, $a = 25$, and $c = 15$.

4. Find the measure of the largest angle of a triangle whose side lengths are 2.9, 3.3, and 4.1.

5. Find the magnitude of the resultant force when forces of 2.6 and 4.3 pounds act on an object with an angle of 40° between them.

6. A ship is supposed to travel directly from city A to city B. The distance between the cities is 20 kilometers. After traveling a distance of 8 kilometers, the captain discovers that the ship has been traveling 18° off course. At this point, how far is the ship from city B?

7. The lengths of two adjacent sides of a parallelogram are 6 centimeters and 8 centimeters. The included angle is 67°. Find the length of the longer diagonal.

8. Three circles with radii of lengths 1, 3, and 4 are tangent to each other. Solve the triangle ABC that connects their centers. (See the following diagram.)

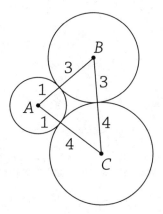

9. Two airplanes leave an airfield at the same time. One flies at an angle of 30° east of due north at 250 kilometers per hour, and the other flies at an angle of 45° east of due south at 300 kilometers per hour. How far apart are the airplanes after 2 hours? (See the following diagram.)

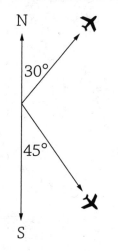

10. Points B and C are at opposite ends of a lake. If point A is 1,500 feet from point B and 2,000 feet from point C and the angle at A is 50°, how long is the lake?

EXERCISE 5-2

Solve as indicated. (Round answers to one decimal place, as needed.)

1. Find a.

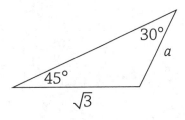

2. Find b.

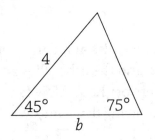

3. Find b.

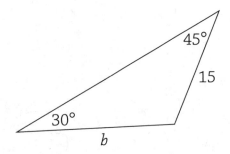

4. Find θ.

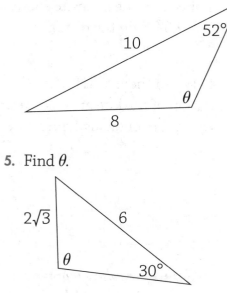

5. Find θ.

6. Find c.

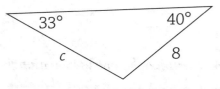

7. Two forces act on an object so that the resultant force is 50 pounds. The measures of the angles between the resultant and the two forces are 25° and 42°. Find the magnitude of the larger applied force to the nearest pound.

8. From two points on shore, the angles A and B from the ground to a light source at C are 16° and 58°, respectively. If AB is 1,500 feet, find the distance, d, of the light source from the shore. (See the following diagram.)

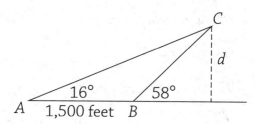

9. Two observers at points A and B are on the same side of a river and are 2,500 feet apart. They both spot an elk at point C on the opposite shore. The angle BAC is 78.5°, and the angle ABC is 47.3°. Find the distance from B to C.

10. Two women 400 feet apart observe a kite between them that is in their vertical plane. The respective angles of elevation of the kite are observed by the women to be 73° and 50.4°, as shown in the following figure. Find the height of the kite above the ground.

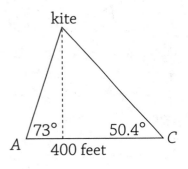

EXERCISE 5-3

For the given information, decide whether there is no triangle, one unique triangle, or two triangles.

1. Given $A = 120°$, $c = 12$, and $a = 16$.

2. Given $A = 38°$, $c = 12$, and $a = 22$.

3. Given $A = 45°$, $c = 12$, and $a = 7$.

4. Given $A = 104°$, $c = 38$, and $a = 22$.

5. Given $A = 15°$, $c = 8$, and $a = 5$.

6. Given $A = 45°$, $c = 38$, and $a = 42$.

7. Given $A = 74°$, $c = 22$, and $a = 22$.

8. Given $A = 60°$, $c = 38$, and $a = 16$.

9. Given $A = 65°$, $c = 38$, and $a = 22$.

10. Given $A = 30°$, $c = 44$, and $a = 22$.

EXERCISE 5-4

For questions 1 to 5 solve triangle ABC for the indicated part. Round answers to one decimal place, as needed.

1. $a = 7.5$, $b = 4.6$, and $c = 7$. Find B.

2. $a = 15$, $A = 25°$, and $B = 40°$. Find b.

3. $a = 315$, $b = 460$, and $A = 42°$. Find B.

4. $a = 17$, $c = 14$, and $B = 30°$. Find b.

5. $c = 190$, $a = 150$, and $C = 85.2°$. Find A.

For questions 6 to 10, solve as indicated. Round answers to one decimal place, as needed.

6. A hot-air balloon is anchored to the ground at point *A* by a rope that is 150 yards long and at point *B* by a rope that is 120 yards long. If the angle between the two ropes is 85°, what is the distance between points *A* and *B*? (See the following diagram.)

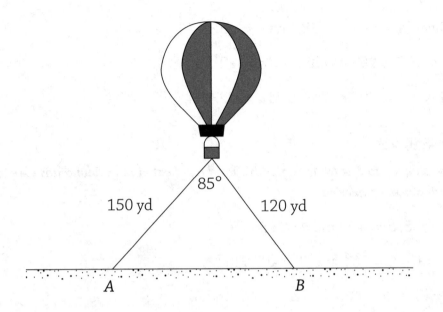

7. A survey line, beginning at point A, must cross a swampy area. From point A, the surveyor sighted a point B at an angle of 52° at a distance of 1,550 feet from point A. At point B, the surveyor turned an angle of 90° and ran a line BC. If C is on the line through the swamp, how far is point C from point B? (See the following diagram.)

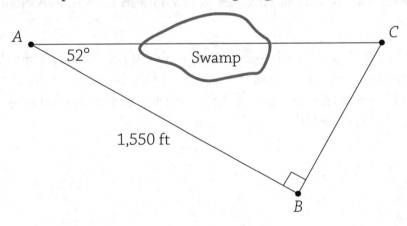

8. Two photographers each take a picture of the same elk drinking at a river's edge. As shown in the following diagram, the two photographers are 400 feet apart, one at point A and one at point B. Find the distance from the photographer at point A to the elk.

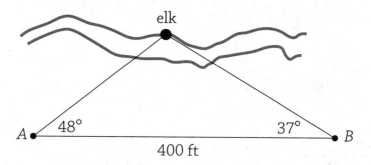

9. Two forces of 18 and 23 pounds act on an object. If the angle between the two forces is 50°, find the magnitude of their resultant. (See the following diagram.)

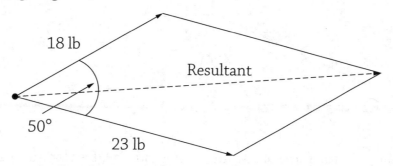

10. The angle of elevation from an observer to a flagpole is 21°. If the observer approaches the flagpole by 24 meters, the angle of elevation becomes 35°. Determine the height of the flagpole.

EXERCISE 5-5

Solve as indicated. (Round answers to one decimal place, as needed).

1. Find the area of the given triangle.

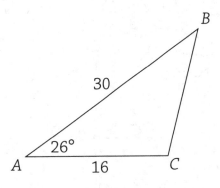

2. Find the area of the given triangle.

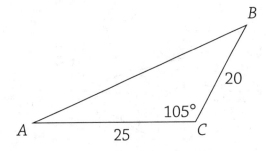

3. A jib sail is in the form of a triangle. One side is 8 feet long and another side is 6 feet long, and the angle included between the sides is 54°. How much material, in ft^2, is contained in the sail?

4. A floor tile is made in the form of an equilateral triangle whose sides measure 10 inches each. What is the area, in in^2, of the tile?

5. A pendant for a necklace is made in the shape of a regular hexagon whose sides measure 3 centimeters each. What is the area, in cm^2, of the pendant?

Flashcard App

6 Trigonometric Functions of Any Angle

MUST KNOW

⚡ The trigonometric relationships between the acute angles and sides of a right triangle can be extended to the concept of trigonometric functions of any angle—including angles greater than 90°, negative angles, and quadrantal angles.

⚡ The unit circle, centered at the origin with radius 1, is a convenient framework for applying trigonometry in the coordinate plane.

e define the trigonometric functions of any angle in terms of coordinates in the *xy*-coordinate plane. Let θ be any angle in standard position in the coordinate plane and (x, y) be a point on θ's terminal side such that the line segment, $r = \sqrt{x^2 + y^2}$, from the origin to (x, y) is not zero (illustrated next).

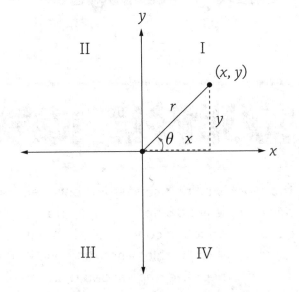

The trigonometric functions of θ are defined as follows:

Definitions of the Trigonometric Functions of Any Angle θ

$\sin \theta = \dfrac{y}{r}$	$\cos \theta = \dfrac{x}{r}$
$\tan \theta = \dfrac{y}{x}, x \neq 0$	$\cot \theta = \dfrac{x}{y}, y \neq 0$
$\sec \theta = \dfrac{r}{x}, x \neq 0$	$\csc \theta = \dfrac{r}{y}, y \neq 0$

BTW

These definitions are well-defined, meaning the values of the trigonometric functions of an angle θ are not dependent on the choice of the point (x, y) selected on the terminal side of θ.

Because $r = \sqrt{x^2 + y^2} \neq 0$, the sine and cosine functions are defined for all real values of θ. The tangent and secant are undefined when $x = 0$; and, similarly, the cotangent and cosecant are undefined when $y = 0$.

EXAMPLE

▶ Using the following figure, find the sine, cosine, and tangent of θ.

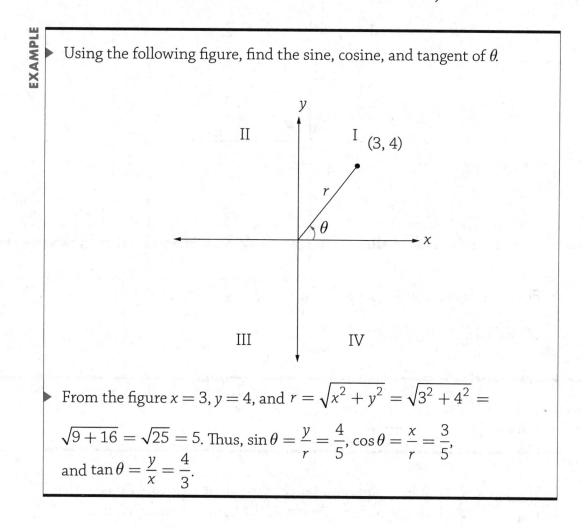

▶ From the figure $x = 3$, $y = 4$, and $r = \sqrt{x^2 + y^2} = \sqrt{3^2 + 4^2} = \sqrt{9 + 16} = \sqrt{25} = 5$. Thus, $\sin\theta = \dfrac{y}{r} = \dfrac{4}{5}$, $\cos\theta = \dfrac{x}{r} = \dfrac{3}{5}$, and $\tan\theta = \dfrac{y}{x} = \dfrac{4}{3}$.

Look at this example where the angle of interest lies in quadrant II.

EXAMPLE

▶ Using the following figure, find the sine, cosine, and tangent of θ.

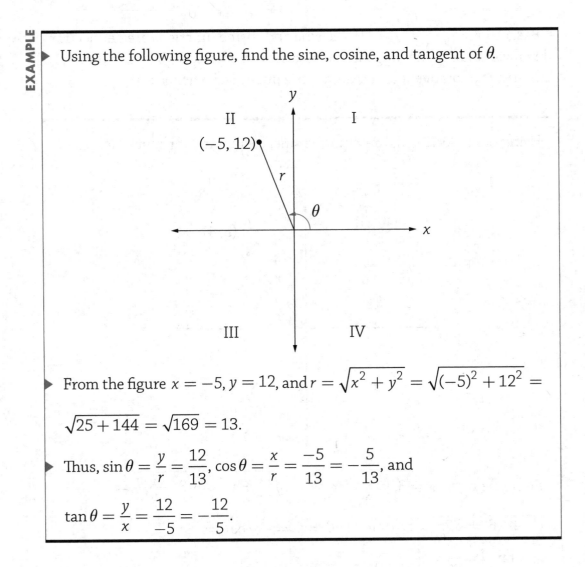

▶ From the figure $x = -5$, $y = 12$, and $r = \sqrt{x^2 + y^2} = \sqrt{(-5)^2 + 12^2} =$

$\sqrt{25 + 144} = \sqrt{169} = 13.$

▶ Thus, $\sin \theta = \dfrac{y}{r} = \dfrac{12}{13}$, $\cos \theta = \dfrac{x}{r} = \dfrac{-5}{13} = -\dfrac{5}{13}$, and

$\tan \theta = \dfrac{y}{x} = \dfrac{12}{-5} = -\dfrac{12}{5}.$

The signs of the sine, cosine, and tangent trigonometric functions in the four quadrants follow the pattern shown in the figure provided next.

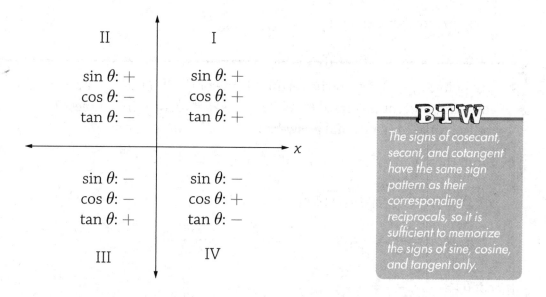

BTW

The signs of cosecant, secant, and cotangent have the same sign pattern as their corresponding reciprocals, so it is sufficient to memorize the signs of sine, cosine, and tangent only.

EXAMPLE

The point $(-2, -2)$ lies on the terminal side of θ. Find the sine, cosine, and tangent of θ. (Write the exact answer in simplest radical form for irrational answers.)

From the given information $x = -2, y = -2$, and
$$r = \sqrt{x^2 + y^2} = \sqrt{(-2)^2 + (-2)^2} = \sqrt{4 + 4} = \sqrt{8} = 2\sqrt{2}.$$

Thus, $\sin\theta = \dfrac{y}{r} = \dfrac{-2}{2\sqrt{2}} = -\dfrac{1}{\sqrt{2}} = -\dfrac{\sqrt{2}}{2}$,

$\cos\theta = \dfrac{x}{r} = \dfrac{-2}{2\sqrt{2}} = -\dfrac{1}{\sqrt{2}} = -\dfrac{\sqrt{2}}{2}$, and $\tan\theta = \dfrac{y}{x} = \dfrac{-2}{-2} = 1$.

Let's try another example.

EXAMPLE

▶ The point $(3, -4)$ lies on the terminal side of θ. Find the secant, cosecant, and cotangent of θ. (Write the exact answer in simplest radical form for irrational answers.)

▶ From the given information $x = 3$, $y = -4$, and $r = \sqrt{x^2 + y^2} =$

$\sqrt{(3)^2 + (-4)^2} = \sqrt{9 + 16} = \sqrt{25} = 5.$

▶ Thus, $\sec\theta = \dfrac{r}{x} = \dfrac{5}{3}$, $\csc\theta = \dfrac{r}{y} = \dfrac{5}{-4} = -\dfrac{5}{4}$, and

$\cot\theta = \dfrac{x}{y} = \dfrac{3}{-4} = -\dfrac{3}{4}.$

In the next example, we strategically assign values to x and r based on the fractional value of the cosine function.

EXAMPLE

▶ Given θ is in quadrant II and $\cos\theta = -\dfrac{4}{5}$, find the exact values of $\sin\theta$ and $\tan\theta$.

▶ Because θ is in quadrant II and $\cos\theta = -\dfrac{4}{5} = \dfrac{x}{r}$, we can let $r = 5$ and

$x = -4$, from which you have $y = \sqrt{5^2 - (-4)^2} = \sqrt{25 - 16} = \sqrt{9} = 3$

(which is positive because θ is in quadrant II).

▶ So, $\sin\theta = \dfrac{y}{r} = \dfrac{3}{5}$ and $\tan\theta = \dfrac{y}{x} = \dfrac{3}{-4} = -\dfrac{3}{4}.$

Trigonometric Functions of Complementary Angles

In right triangle ABC shown below, with $C = 90°$, the acute angles α and β are complementary.

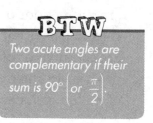

BTW

Two acute angles are complementary if their sum is $90°$ $\left(\text{or } \dfrac{\pi}{2}\right)$.

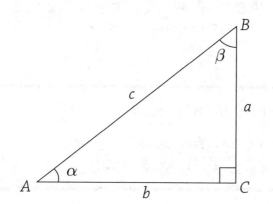

From the figure, you have the following **cofunction** relationships:

Cofunction Relationships of Complementary Angles

$$\sin\beta = \cos(90° - \beta) = \cos\left(\frac{\pi}{2} - \beta\right) = \frac{b}{c} = \cos\alpha$$

$$\cos\beta = \sin(90° - \beta) = \sin\left(\frac{\pi}{2} - \beta\right) = \frac{a}{c} = \sin\alpha$$

$$\tan\beta = \cot(90° - \beta) = \cot\left(\frac{\pi}{2} - \beta\right) = \frac{b}{a} = \cot\alpha$$

$$\sec\beta = \csc(90° - \beta) = \csc\left(\frac{\pi}{2} - \beta\right) = \frac{c}{a} = \csc\alpha$$

$$\csc\beta = \sec(90° - \beta) = \sec\left(\frac{\pi}{2} - \beta\right) = \frac{c}{b} = \sec\alpha$$

$$\cot\beta = \tan(90° - \beta) = \tan\left(\frac{\pi}{2} - \beta\right) = \frac{a}{b} = \tan\alpha$$

These relationships occur in pairs and are valid for any pair of complementary angles. Each function of a pair is the cofunction of the other: sine and cosine are cofunctions of each other, secant and cosecant are cofunctions of each other, and tangent and cotangent are cofunctions of each other. In general, any function of an acute angle equals the corresponding cofunction of its complementary angle.

The cofunction relationships are the basis for why the functions cosine, cosecant, and cotangent have "co" in their names.

EXAMPLE

▶ Find an acute angle θ such that $\sin 28° = \cos \theta$.

▶ We know that sine and cosine are cofunctions. Therefore, $\sin 28° = \cos(90° - 28°) = \cos 62°$. Hence, $\theta = 62°$.

Knowing the cofunction relationships can be helpful in solving certain trigonometric equations.

EXAMPLE

▶ Solve $\sin(2\theta + 5°) = \cos(33°)$, where θ is an acute angle.

▶ Because sine and cosine are cofunctions, $2\theta + 5°$ and $33°$ are complementary angles.

▶ Therefore,

$$2\theta + 5° = 90° - 33°$$
$$2\theta + 5° = 57°$$
$$2\theta = 52°$$
$$\theta = 26°$$

Unit Circle Trigonometry

The **unit circle** is the circle, centered at the origin, with radius 1 consisting of all points that satisfy the equation $x^2 + y^2 = 1$. It is a convenient framework for applying trigonometry in the coordinate plane. If θ is the central angle in standard position that is formed by the line segment from the origin to a point (x, y) on the unit circle, then, by definition of sine and cosine, $(x, y) = (\cos \theta, \sin \theta)$. (See the following illustration.)

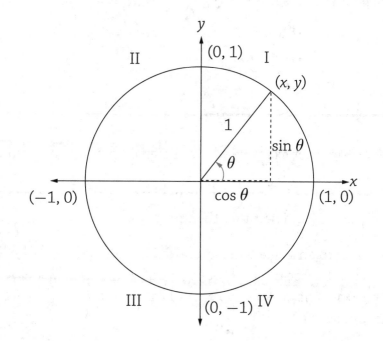

From the unit circle, given that $x = \cos \theta$ and $y = \sin \theta$, you can obtain the other trigonometric functions for θ as follows:

$$\tan \theta = \frac{y}{x} = \frac{\sin \theta}{\cos \theta}, \cot \theta = \frac{x}{y} = \frac{\cos \theta}{\sin \theta},$$

$$\sec \theta = \frac{1}{x} = \frac{1}{\cos \theta} \text{ and } \csc \theta = \frac{1}{y} = \frac{1}{\sin \theta}$$

BTW

It is important to be aware that, for any given angle, the value of a trigonometric function remains the same, regardless of the radius of the circle.

Let's look at an example where the given point is in quadrant III.

EXAMPLE

▶ The point $\left(-\dfrac{\sqrt{3}}{2}, -\dfrac{1}{2}\right)$ is a point on the unit circle corresponding to an angle θ in standard position. Find the sine, cosine, and tangent of θ. (Write the exact answer in simplest radical form for irrational answers.)

▶ From the given information, $\sin\theta = -\dfrac{1}{2}$, $\cos\theta = -\dfrac{\sqrt{3}}{2}$, and

$$\tan\theta = \frac{\sin\theta}{\cos\theta} = \frac{-\dfrac{\sqrt{3}}{2}}{-\dfrac{1}{2}} = \sqrt{3}.$$

Now let's try another example where the given point is in quadrant IV.

EXAMPLE

▶ The point $\left(\dfrac{4}{5}, -\dfrac{3}{5}\right)$ is a point on the unit circle corresponding to an angle θ in standard position. Find the secant, cosecant, and cotangent of θ. (Write the exact answer in simplest radical form for irrational answers.)

▶ From the given information, $\sin\theta = -\dfrac{3}{5}$ and $\cos\theta = \dfrac{4}{5}$. Then, $\sec\theta = \dfrac{1}{\cos\theta} = \dfrac{1}{\left(\dfrac{4}{5}\right)} = \dfrac{5}{4}$,

$\csc\theta = \dfrac{1}{\sin\theta} = \dfrac{1}{\left(-\dfrac{3}{5}\right)} = -\dfrac{5}{3}$, and $\cot\theta = \dfrac{\cos\theta}{\sin\theta} = \dfrac{\left(\dfrac{4}{5}\right)}{\left(-\dfrac{3}{5}\right)} = -\dfrac{4}{3}$.

BTW

The unit circle has its historical beginnings in the works of Greek mathematicians who made use of chords of circles. The Greek mathematician Hipparchus produced the first known table of chords around 140 BCE.

Trigonometric Functions of Quadrantal Angles

A **quadrantal angle** is an angle in standard position whose terminal side lies on an axis. The angles 0°, 90°, 180°, and 270° and all the angles coterminal with them are quadrantal angles.

Some examples of quadrantal angles are

■ 0°, 90°, 180°, 270°, 360°, 450°, and so on

■ −90°, −180°, −270°, −360°, −450°, and so on

■ $0, \dfrac{\pi}{2}, \pi, \dfrac{3\pi}{2}, 2\pi, \dfrac{5\pi}{2}$, and so on

■ $-\dfrac{\pi}{2}, -\pi, -\dfrac{3\pi}{2}, -2\pi, -\dfrac{5\pi}{2}$, and so on

> **BTW**
>
> *Quadrantal angles are multiples of 90° $\left(\text{or } \dfrac{\pi}{2} \right)$.*

Using the unit circle and the definitions of the trigonometric functions, we can determine the trigonometric function values for quadrantal angles (see the following figure).

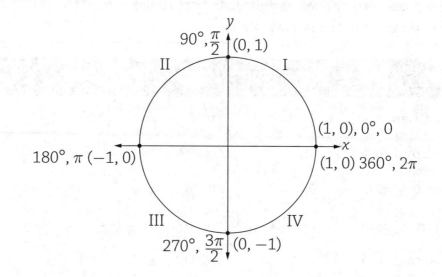

▶ Given the point $(x, y) = (0, 1)$ on the unit circle corresponding to an angle of $90°$ in standard position, find the values of the six trigonometric functions for an angle of $90°$.

▶ From the information given, $x = 0$ and $y = 1$. We know that $x = \cos 90°$ and $y = \sin 90°$.

So, $\cos 90° = 0$ and $\sin 90° = 1$.

▶ Then, $\tan 90° = \dfrac{\sin 90°}{\cos 90°} = \dfrac{1}{0} = \text{undefined}$,

$\sec 90° = \dfrac{1}{0} = \text{undefined}$, $\csc 90° = \dfrac{1}{\sin 90°} =$

$\dfrac{1}{1} = 1$, and $\cot 90° = \dfrac{\cos 90°}{\sin 90°} = \dfrac{0}{1} = 0$.

BTW

The use of the "=" sign in "= undefined" to indicate that a trigonometric function is undefined does not have the customary meaning of "equals." The notation "= undefined" is used in math as a short way to describe a special situation in which an expression has no meaning.

The trigonometric values of the quadrantal angles from $0°$ to $360°$ (0 to 2π) are given in the following table.

$\theta = 0° = 0$; $(x, y) = (1, 0)$, $r = 1$	$\theta = 90° = \dfrac{\pi}{2}$; $(x, y) = (0, 1)$, $r = 1$
$\sin 0° = \sin 0 = \dfrac{0}{1} = 0$	$\sin 90° = \sin\dfrac{\pi}{2} = \dfrac{1}{1} = 1$
$\cos 0° = \cos 0 = \dfrac{1}{1} = 1$	$\cos 90° = \cos\dfrac{\pi}{2} = \dfrac{0}{1} = 0$
$\tan 0° = \tan 0 = \dfrac{0}{1} = 0$	$\tan 90° = \tan\dfrac{\pi}{2} = \dfrac{1}{0} = \text{undefined}$
$\sec 0° = \sec 0 = \dfrac{1}{1} = 1$	$\sec 90° = \sec\dfrac{\pi}{2} = \dfrac{1}{0} = \text{undefined}$
$\csc 0° = \csc 0 = \dfrac{1}{0} = \text{undefined}$	$\csc 90° = \csc\dfrac{\pi}{2} = \dfrac{1}{1} = 1$
$\cot 0° = \cot 0 = \dfrac{1}{0} = \text{undefined}$	$\cot 90° = \cot\dfrac{\pi}{2} = \dfrac{0}{1} = 0$

$\theta = 180° = \pi; (x,y) = (-1,0), r = 1$	$\theta = 270° = \dfrac{3\pi}{2}; (x,y) = (0,-1), r = 1$
$\sin 180° = \sin \pi = \dfrac{0}{1} = 0$	$\sin 270° = \sin \dfrac{3\pi}{2} = \dfrac{-1}{1} = -1$
$\cos 180° = \cos \pi = \dfrac{-1}{1} = -1$	$\cos 270° = \cos \dfrac{3\pi}{2} = \dfrac{0}{1} = 0$
$\tan 180° = \tan \pi = \dfrac{0}{-1} = 0$	$\tan 270° = \tan \dfrac{3\pi}{2} = \dfrac{-1}{0} = $ undefined
$\sec 180° = \sec \pi = \dfrac{1}{-1} = -1$	$\sec 270° = \sec \dfrac{3\pi}{2} = \dfrac{1}{0} = $ undefined
$\csc 180° = \csc \pi = \dfrac{1}{0} = $ undefined	$\csc 270° = \csc \dfrac{3\pi}{2} = \dfrac{1}{-1} = -1$
$\cot 180° = \cot \pi = \dfrac{-1}{0} = $ undefined	$\cot 270° = \cot \dfrac{3\pi}{2} = \dfrac{0}{-1} = 0$

We can use what we know about quadrantal angles to evaluate trigonometric expressions that contain quadrantal angles.

EXAMPLE

Evaluate $\cos 0° - 4\sin 270° + 2\cot 90°$.

$\cos 0° - 4\sin 270° + 2\cot 90° = 1 - 4(-1) + 2(0) = 1 + 4 + 0 = 5$

The next example makes use of the notation $\cos^2 \pi$, which is a special way to write $(\cos \pi)^2$.

EXAMPLE

Evaluate $-3\cos^2 \pi + 2\csc \dfrac{3\pi}{2} + \tan 0$.

$-3\cos^2 \pi + 2\csc \dfrac{3\pi}{2} + \tan 0 = -3(-1)^2 + 2(-1) + (0) = -3 - 2 + 0 = -5$

Trigonometric Functions of Coterminal Angles

The values of the trigonometric functions remain the same if an angle is replaced by one that is coterminal with the angle (refer back to Chapter 1 for a thorough discussion of coterminal angles). If an angle is greater than $360°$ (or 2π) or is negative, you can find an equivalent nonnegative coterminal angle that is less than $360°$ (or 2π) by adding or subtracting a positive integer multiple of $360°$ (or 2π).

Let's try an example in which the angle is expressed in degrees.

EXAMPLE

▶ Find the exact value of $\cos 1140°$.

▶ Because $1140° - (3 \times 360°) = 1140° - 1080° = 60°$,

$$\cos 1140° = \cos 60° = \frac{1}{2}.$$

Now let's do one in which the angle is expressed in radians.

EXAMPLE

▶ Find the exact value of $\sin\left(-\dfrac{7\pi}{2}\right)$.

▶ Because $-\dfrac{7\pi}{2} + (2 \times 2\pi) = -\dfrac{7\pi}{2} + 4\pi = \dfrac{\pi}{2}$, $\sin\left(-\dfrac{7\pi}{2}\right) = \sin\dfrac{\pi}{2} = 1$.

Trigonometric Functions of Negative Angles

Given an angle θ, trigonometric functions for which the trigonometric value of $-\theta$ equals the trigonometric value of θ are **even functions**, and those for which the trigonometric value of $-\theta$ equals the negative of the trigonometric value of θ are **odd functions**. As shown in the following table, cosine and secant are even trigonometric functions, and sine, cosecant, tangent, and cotangent are odd trigonometric functions.

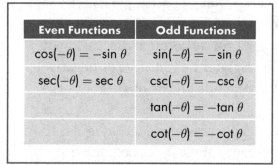

Even Functions	Odd Functions
$\cos(-\theta) = -\sin\theta$	$\sin(-\theta) = -\sin\theta$
$\sec(-\theta) = \sec\theta$	$\csc(-\theta) = -\csc\theta$
	$\tan(-\theta) = -\tan\theta$
	$\cot(-\theta) = -\cot\theta$

Applying the information in the table, we have the following results:

- $\sin(-75°) = -\sin 75°$

- $\tan\left(-\dfrac{\pi}{5}\right) = -\tan\dfrac{\pi}{5}$

- $\cos(-250°) = \cos 250°$

- If $\sin(-\theta) = \dfrac{4}{5}$, then $\sin\theta = -\dfrac{4}{5}$.

- If $\sec(-\theta) = 2$, then $\cos\theta = \dfrac{1}{2}$.

Let's try an example involving a special angle.

EXAMPLE

▶ Find the exact value of $\tan 330°$.

▶ Because $330°$ and $-30°$ are coterminal, $\tan 330° = \tan(-30°)$. And because the tangent function is odd, $\tan(-30°) = -\tan 30° = -\dfrac{\sqrt{3}}{3}$.

▶ Thus, $\tan 330° = \tan(-30°) = -\tan 30° = -\dfrac{\sqrt{3}}{2}$.

Using Reference Angles to Find the Values of Trigonometric Functions

A trigonometric function value of an angle has the same absolute value as the trigonometric function value of its reference angle (refer to Chapter 1 for a thorough discussion of reference angles). The sign (either positive or negative) of the function value depends on the quadrant of the original angle.

> **BTW**
>
> *If θ is a quadrantal angle, then the values of the trigonometric functions for θ are the same as given earlier in this chapter, so a reference angle is not needed.*

The relationship of a positive angle θ that is less than $360°$ (2π) and its reference angle θ_r in each quadrant is given in the following table.

	θ's Quadrant	θ_r Signs of Functions
I	$\theta_r = \theta$	All functions are positive.
II	$\theta_r = 180° - \theta$ or $\pi - \theta$	Only $\sin \theta$ and $\csc \theta$ are positive.
III	$\theta_r = \theta - 180°$ or $\theta - \pi$	Only $\tan \theta$ and $\cot \theta$ are positive.
IV	$\theta_r = 360° - \theta$ or $2\pi - \theta$	Only $\cos \theta$ and $\sec \theta$ are positive.

Let's try an example where the given angle is in quadrant III.

EXAMPLE

▶ Express $\sin 235°$ as the sine of a positive acute angle.

▶ Because $235°$ is in quadrant III, we know its sine is negative and that we obtain its reference angle by using $\theta_r = \theta - 180°$.

▶ Thus, $\sin 235° = -\sin(235° - 180°) = -\sin 55°$.

Now, for an example where the given angle is in quadrant II.

EXAMPLE

Express $\tan \dfrac{3\pi}{5}$ as the tangent of a positive acute angle.

Because $\dfrac{3\pi}{5}$ is in quadrant II, we know its tangent is negative and that we obtain its reference angle by using $\theta_r = \pi - \theta$.

Thus, $\tan \dfrac{3\pi}{5} = -\tan\left(\pi - \dfrac{3\pi}{5}\right) = -\tan \dfrac{2\pi}{5}$.

When the reference angle for a trigonometric function is a special acute angle, you can determine the exact value of the trigonometric function. Refer to Chapter 3 for a thorough discussion of special acute angles.

EXAMPLE

Find the exact value of $\sin \dfrac{5\pi}{6}$.

Because $\dfrac{5\pi}{6}$ is in quadrant II, we know its sine is positive and that we obtain its reference angle by using $\theta_r = \pi - \theta$.

Thus, $\sin \dfrac{5\pi}{6} = \sin\left(\pi - \dfrac{5\pi}{6}\right) = \sin \dfrac{\pi}{6} = \dfrac{1}{2}$.

For your convenience, values of the trigonometric functions for special angles and the quadrantal angles from $0°$ to $360°$ (0 to 2π) are given in the following table.

Trigonometric Function Values of Special Angles and Quadrantal Angles						
θ (Deg or Rad)	$\sin\theta$	$\cos\theta$	$\tan\theta$	$\sec\theta$	$\csc\theta$	$\cot\theta$
$\theta°$ or 0	0	1	0	1	undefined	undefined
30° or $\dfrac{\pi}{6}$	$\dfrac{1}{2}$	$\dfrac{\sqrt{3}}{2}$	$\dfrac{\sqrt{3}}{3}$	$\dfrac{2\sqrt{3}}{3}$	2	$\sqrt{3}$
45° or $\dfrac{\pi}{4}$	$\dfrac{\sqrt{2}}{2}$	$\dfrac{\sqrt{2}}{2}$	1	$\sqrt{2}$	$\sqrt{2}$	1
60° or $\dfrac{\pi}{3}$	$\dfrac{\sqrt{3}}{2}$	$\dfrac{1}{2}$	$\sqrt{3}$	2	$\dfrac{2\sqrt{3}}{3}$	$\dfrac{\sqrt{3}}{3}$
90° or $\dfrac{\pi}{2}$	1	0	undefined	undefined	1	0
120° or $\dfrac{2\pi}{3}$	$\dfrac{\sqrt{3}}{2}$	$-\dfrac{1}{2}$	$-\sqrt{3}$	-2	$\dfrac{2\sqrt{3}}{3}$	$-\dfrac{\sqrt{3}}{3}$
135° or $\dfrac{3\pi}{4}$	$\dfrac{\sqrt{2}}{2}$	$-\dfrac{\sqrt{2}}{2}$	-1	$-\sqrt{2}$	$\sqrt{2}$	-1
150° or $\dfrac{5\pi}{6}$	$\dfrac{1}{2}$	$-\dfrac{\sqrt{3}}{2}$	$-\dfrac{\sqrt{3}}{3}$	$-\dfrac{2\sqrt{3}}{3}$	2	$-\sqrt{3}$
180° or π	0	-1	0	-1	undefined	undefined
210° or $\dfrac{7\pi}{6}$	$-\dfrac{1}{2}$	$-\dfrac{\sqrt{3}}{2}$	$\dfrac{\sqrt{3}}{3}$	$-\dfrac{2\sqrt{3}}{3}$	-2	$\sqrt{3}$
225° or $\dfrac{5\pi}{4}$	$-\dfrac{\sqrt{2}}{2}$	$-\dfrac{\sqrt{2}}{2}$	1	$-\sqrt{2}$	$-\sqrt{2}$	1
240° or $\dfrac{4\pi}{3}$	$-\dfrac{\sqrt{3}}{2}$	$-\dfrac{1}{2}$	$\sqrt{3}$	-2	$-\dfrac{2\sqrt{3}}{3}$	$\dfrac{\sqrt{3}}{3}$
270° or $\dfrac{3\pi}{2}$	-1	0	undefined	undefined	-1	0
300° or $\dfrac{5\pi}{3}$	$-\dfrac{\sqrt{3}}{2}$	$\dfrac{1}{2}$	$-\sqrt{3}$	2	$-\dfrac{2\sqrt{3}}{3}$	$-\dfrac{\sqrt{3}}{3}$
315° or $\dfrac{7\pi}{4}$	$-\dfrac{\sqrt{2}}{2}$	$\dfrac{\sqrt{2}}{2}$	-1	$\sqrt{2}$	$-\sqrt{2}$	-1
330° or $\dfrac{11\pi}{6}$	$-\dfrac{1}{2}$	$\dfrac{\sqrt{3}}{2}$	$-\dfrac{\sqrt{3}}{3}$	$\dfrac{2\sqrt{3}}{3}$	-2	$-\sqrt{3}$
360° or 2π	0	1	0	1	undefined	undefined

We frequently encounter these angles in the study of trigonometry. Given your understanding of reference angles, if you know the function values of 0°, 30°, 45°, 60°, and 90° and you know the signs of the trigonometric functions in each of the four quadrants, you can reproduce the table shown here without the use of any other aid. For most other angles, a calculator (or similar resource) is needed to determine the values of their trigonometric functions.

BTW

The algebraic sign of the three basic trigonometric functions in each quadrant is easily remembered by using the mnemonic "All students take calculus," which reminds you that "All three are positive in quadrant I, the sine is positive in quadrant II, the tangent is positive in quadrant III, and the cosine is positive in quadrant IV."

EXERCISES

EXERCISE 6-1

For questions 1 to 5, solve as indicated. (Write the exact answer in simplest radical form for irrational answers.)

1. The point $(2\sqrt{6}, -1)$ lies on the terminal side of θ. Find the sine, cosine, and tangent of θ.

2. The point $(-3, \sqrt{7})$ lies on the terminal side of θ. Find the sine, cosine, and tangent of θ.

3. The point $(-8, -15)$ lies on the terminal side of θ. Find the sine, cosine, and tangent of θ.

4. The point $(-12, 5)$ lies on the terminal side of θ. Find the secant, cosecant, and cotangent of θ.

5. The point $(7, 24)$ lies on the terminal side of θ. Find the secant, cosecant, and cotangent of θ.

For questions 6 to 10, for the given information, name the quadrant in which θ lies.

6. $\tan\theta > 0$ and $\cos\theta < 0$

7. $\sin\theta > 0$ and $\sec\theta < 0$

8. $\csc\theta > 0$ and $\cos\theta > 0$

9. $\sec\theta < 0$ and $\cot\theta > 0$

10. $\sin\theta > 0$ and $\cot\theta < 0$

For questions 11 to 15, solve as indicated. (Write the exact answer in simplest radical form for irrational answers.)

11. Given θ is in quadrant III and $\cos\theta = -\dfrac{5}{13}$, find the exact values of $\sin\theta$ and $\tan\theta$.

12. Given θ is in quadrant IV and $\sin \theta = -\dfrac{6}{10}$, find the exact values of $\cos \theta$ and $\tan \theta$.

13. Given θ is in quadrant I and $\sec \theta = \dfrac{41}{9}$, find the exact values of $\sin \theta$ and $\cos \theta$.

14. Given θ is in quadrant II and $\tan \theta = -\dfrac{7}{24}$, find the exact values of $\sin \theta$ and $\cos \theta$.

15. Given θ is in quadrant III and $\sin \theta = -\dfrac{15}{17}$, find the exact values of $\cos \theta$ and $\tan \theta$.

EXERCISE 6-2

Solve for θ, where θ is an acute angle.

1. $\cos 54° = \sin \theta$

2. $\tan \dfrac{\pi}{3} = \cot \theta$

3. $\sin 48° = \cos(\theta + 7°)$

4. $\tan 63° = \cot(\theta - 40°)$

5. $\sec\left(5\theta + \dfrac{\pi}{12}\right) = \csc\left(3\theta - \dfrac{\pi}{4}\right)$

EXERCISE 6-3

Solve as indicated. (Write the exact answer in simplest radical form for irrational answers.)

1. The point $\left(\dfrac{1}{2}, \dfrac{\sqrt{3}}{2}\right)$ is a point on the unit circle corresponding to an angle θ in standard position. Find the sine, cosine, and tangent of θ.

2. The point $\left(\dfrac{\sqrt{3}}{2}, -\dfrac{1}{2}\right)$ is a point on the unit circle corresponding to an angle θ in standard position. Find the sine, cosine, and tangent of θ.

3. The point $\left(-\dfrac{\sqrt{2}}{2}, \dfrac{\sqrt{2}}{2}\right)$ is a point on the unit circle corresponding to an angle θ in standard position. Find the sine, cosine, and tangent of θ.

4. The point $\left(-\dfrac{1}{2}, -\dfrac{\sqrt{3}}{2}\right)$ is a point on the unit circle corresponding to an angle θ in standard position. Find the sine, cosine, and tangent of θ.

5. The point $\left(-\dfrac{12}{13}, \dfrac{5}{13}\right)$ is a point on the unit circle corresponding to an angle θ in standard position. Find the sine, cosine, and tangent of θ.

6. The point $\left(-\dfrac{9}{41}, -\dfrac{40}{41}\right)$ is a point on the unit circle corresponding to an angle θ in standard position. Find the secant, cosecant, and cotangent of θ.

7. The point $\left(\dfrac{\sqrt{2}}{2}, -\dfrac{\sqrt{2}}{2}\right)$ is a point on the unit circle corresponding to an angle θ in standard position. Find the secant, cosecant, and cotangent of θ.

8. The point $\left(-\dfrac{12}{13}, \dfrac{5}{13}\right)$ is a point on the unit circle corresponding to an angle θ in standard position. Find the secant, cosecant, and cotangent of θ.

9. The point $\left(-\dfrac{7}{25}, \dfrac{24}{25}\right)$ is a point on the unit circle corresponding to an angle θ in standard position. Find the secant, cosecant, and cotangent of θ.

10. The point $\left(\dfrac{36}{85}, -\dfrac{77}{85}\right)$ is a point on the unit circle corresponding to an angle θ in standard position. Find the secant, cosecant, and cotangent of θ.

EXERCISE 6-4

Evaluate as indicated.

1. $\csc\dfrac{\pi}{2} - \sin\dfrac{3\pi}{2}$

2. $5\sec^2\pi + 2\tan\pi - \sec\pi$

3. $4\sin^2 90° - \tan 180° + 3\cos^2 90°$

4. $\tan\pi - 2\cos\pi + 3\csc\dfrac{3\pi}{2} + \sin\dfrac{\pi}{2}$

5. $4\cos 270° - 5\sec 180° - 6\csc 270° + \sin 180° + 2\cos 180° + 3\sin 270°$

EXERCISE 6-5

Replace the given angle(s) with a nonnegative coterminal that is less than 360° or 2π, and then find the exact value of the trigonometric expression.

1. $\cos(-1410°)$

2. $\sec 1485°$

3. $\sin\left(-\dfrac{5\pi}{3}\right)$

4. $\csc 450°$

5. $\tan\dfrac{13\pi}{4}$

6. $5\sqrt{3}\tan\left(-\dfrac{11\pi}{6}\right)$

7. $6\sin 750° - 2\cos 780°$

8. $3\sqrt{2}\sin(-675°) + 2\sqrt{3}\cos(-690°)$

9. $-\tan\dfrac{9\pi}{4}\sin\dfrac{13\pi}{6}$

10. $\cos\left(-\dfrac{17\pi}{3}\right)\sin\left(-\dfrac{17\pi}{3}\right)$

EXERCISE 6-6

For questions 1 to 5, solve as indicated.

1. Given $\tan(-\theta) = \dfrac{1}{3}$. Find $\cot\theta$.

2. Given $\sin(-\theta) = \dfrac{11}{60}$. Find $\csc\theta$.

3. Given $\sec(-\theta) = -\dfrac{25}{7}$. Find $\cos\theta$.

4. Given $\cot(-\theta) = -\dfrac{3}{4}$. Find $\tan\theta$.

5. Given $\cos(-\theta) = \dfrac{48}{73}$. Find $\sec\theta$.

For questions 6 to 10, find the exact value of the given expression.

6. $\cos 300°$

7. $\sin\dfrac{11\pi}{6}$

8. $\sin(-750°)$

9. $\cot\dfrac{7\pi}{4}$

10. $5\sqrt{2}\tan(-45°) + 4\cos(-60°)$

EXERCISE 6-7

For questions 1 to 10, find the exact value of the given expression.

1. $\sin\dfrac{\pi}{2}$

2. $\tan\dfrac{5\pi}{6}$

3. $\sec\pi$

4. $\cot 120°$

5. $\sin(-315°)$

6. $\sec 135°$

7. $\cos\dfrac{3\pi}{4} + \sin\left(-\dfrac{\pi}{4}\right)$

8. $\tan 60° - \cot(-30°)$

9. $\sin\dfrac{3\pi}{4}\cos\dfrac{5\pi}{3}$

10. $\tan 150° \tan 225°$

Flashcard
App

7 Trigonometric Identities

MUST KNOW

- A trigonometric identity is an equation involving one or more trigonometric functions that is true for all values of the variable(s) for which both sides of the equation are defined.

- Using knowledge of trigonometric relationships combined with skills of algebraic manipulation, we verify an identity by transforming one side of the identity until it is identical to the other side.

- Trigonometric identities are crucially important for simplifying complicated trigonometric expressions or equations.

- Some basic identities, such as the reciprocal, ratio, and Pythagorean identities, are used frequently.

To be successful in trigonometry, you need to know and be able to use adroitly the most well-known trigonometric identities. The good news is that you don't have to know every single trigonometric identity by heart, but there are a few for which it is very important that you make them your own. These basic identities are presented in this chapter.

Definition and Guidelines

A **trigonometric identity** is an equation that is true for all values of the variable(s) for which both sides of the equation are defined. (See Appendix B for a list of identities.) Naturally, an underlying assumption for an identity is that no expression in any denominator equals 0. Otherwise, the expression would be undefined.

Consider this example.

EXAMPLE

$\cos \theta \tan \theta = \sin \theta$ is a trigonometric identity because it is true for all values of θ for which both sides of the equation are defined.

We verify an identity by transforming one side of the identity until it is identical to the other side. Here are some guidelines to keep in mind:

- Do not treat the identity as an equation. You cannot assume it is true, so equality properties cannot be applied.

- Work with only one side of the identity. It doesn't matter which side you start with, but as a general rule, start with the side that appears to be more complicated.

- Proceed through a series of steps in which every intermediate step is equivalent to the previous step.

- The proof is finished when the side you started with looks the same as the other side.

- Use appropriate trigonometric identities (such as the ones we will encounter in subsequent lessons in this chapter) and algebraic manipulation (such as factoring, combining fractional expressions, multiplying/squaring expressions, and so forth) to reach the desired result.

- In some cases, converting all trigonometric functions to sines and cosines is helpful.

- Be persistent. Verifying identities is a skill that is learned through lots of practice.

Note that first two bullets in the guidelines are nonnegotiable. However, within those constraints, there is often room for flexibility when you are verifying identities. When looking at the examples and working through the identities that follow, keep in mind there can be multiple ways to transform one side of an identity so that it is identical to the other side. You might think of a way to accomplish that goal other than what is shown.

Let's give it a try.

*Proving or verifying an identity **cannot** be accomplished by repeatedly substituting in domain values to obtain a true equation. Proving identities can be accomplished by the use of good logic, referencing fundamental identities, and using algebraic skills and substitution principles.*

EXAMPLE

▶ Verify the identity: $\cos\theta\tan\theta = \sin\theta$.

▶ The approach is to logically transform one side of the identity into the other side. Starting with the left side, proceed as follows:

$$\cos\theta\tan\theta = \cos\theta\frac{\sin\theta}{\cos\theta} \quad \text{(by the ratio identity for tangent given later in this chapter)}$$

$$= \cancel{\cos\theta}\frac{\sin\theta}{\cancel{\cos\theta}}$$

$$= \sin\theta$$

▶ Therefore, $\cos\theta\tan\theta = \sin\theta$ is an identity.

Let's see what to do when the statement given is not an identity.

EXAMPLE

▶ Verify that $\sin(\theta + \beta) = \sin\theta + \sin\beta$ is *not* an identity.

▶ It takes only one counterexample to establish that a statement is not an identity. For that purpose, we are permitted to substitute in domain values to show the statement is false.

▶ Let $\theta = \beta = \dfrac{\pi}{3}$. Then, on the left side, we have

$$\sin(\theta + \beta) = \sin\left(\dfrac{2\pi}{3}\right) = \dfrac{\sqrt{3}}{2}.$$ But on the right side,

$$\sin\theta + \sin\beta = \sin\dfrac{\pi}{3} + \sin\dfrac{\pi}{3} = \dfrac{\sqrt{3}}{2} + \dfrac{\sqrt{3}}{2} = \sqrt{3}.$$

Because $\dfrac{\sqrt{3}}{2} \neq \sqrt{3}$, the equation is not an identity.

BTW

A *counterexample* is an example that disproves a statement. We can prove that a statement is not true by substituting in a domain value that results in a false statement.

The Reciprocal and Ratio Identities

The **reciprocal identities** are the following:

$$\sec\theta = \dfrac{1}{\cos\theta} \qquad \csc\theta = \dfrac{1}{\sin\theta} \qquad \cot\theta = \dfrac{1}{\tan\theta}$$

Here is an example that uses a reciprocal identity in verifying an identity.

EXAMPLE

▶ Verify the identity: $\sec\theta(\cos\theta + 1) = 1 + \sec\theta$.

▶ Starting with the left side, we proceed as follows:

$$\sec\theta(\cos\theta + 1) = \dfrac{1}{\cos\theta}(\cos\theta + 1)$$

$$= \dfrac{\cos\theta}{\cos\theta} + \dfrac{1}{\cos\theta}$$

$$= 1 + \sec\theta$$

▶ Therefore, $\sec\theta(\cos\theta + 1) = 1 + \sec\theta$ is an identity.

The **ratio identities** are the following:

$$\tan \theta = \frac{\sin \theta}{\cos \theta} \qquad \cot \theta = \frac{\cos \theta}{\sin \theta}$$

Every *trigonometric function can be written in terms of sine and/or cosine functions.*

In the next example, we're going to use a ratio identity to transform an expression containing two trigonometric functions to one containing only one trigonometric function.

EXAMPLE

▶ Use a ratio identity to express $\dfrac{\sin^4 \theta}{\cos^4 \theta}$ in terms of one single trigonometric function.

▶ $$\frac{\sin^4 \theta}{\cos^4 \theta} = \left(\frac{\sin \theta}{\cos \theta} \right)^4 = \tan^4 \theta$$

 IRL   Writing an expression in terms of one single function can be useful for numerical work in the real world. To get a numerical value requires evaluating only one function rather than two or more individually.

The Cofunction, Periodic, and Even-Odd Identities

The **cofunction identities** are the following:

$$\sin\left(\frac{\pi}{2} - \theta\right) = \cos \theta \qquad \cos\left(\frac{\pi}{2} - \theta\right) = \sin \theta$$

$$\tan\left(\frac{\pi}{2} - \theta\right) = \cot \theta$$

Here's an example where the given angle is expressed in degrees.

Throughout the discussions in this book, as applicable, the radian measure of angles can be replaced with their corresponding degree measure without loss of generality. For instance, we are at liberty to express the cofunction identities as sin(90° − θ) = cos θ, cos(90° − θ) = sin θ, and tan(90° − θ) = cot θ, if needed.

EXAMPLE

Determine the acute angle θ using cofunction identities.

$$\sin \theta = \cos 35°$$

Because sine and cosine are cofunctions, $\theta = 90° - 35° = 55°$.

Here's an example where the given angle is expressed in radians.

EXAMPLE

Determine the acute angle θ using cofunction identities.

$$\tan \frac{7\pi}{60} = \cot \theta$$

Because tangent and cotangent are cofunctions, $\theta = \dfrac{\pi}{2} - \dfrac{7\pi}{60} = \dfrac{30\pi}{60} - \dfrac{7\pi}{60} = \dfrac{23\pi}{60}$.

The **periodic identities** are the following:

For any integer n,

$$\sin(\theta \pm n \cdot 2\pi) = \sin \theta \quad \cos(\theta \pm n \cdot 2\pi) = \theta \quad \tan(\theta \pm n \cdot \pi) = \tan \theta$$

These identities are called "periodic" because they specify that the sine and cosine functions repeat their function values in regular intervals of 2π (or 360°), and the tangent function repeats its function values in regular intervals of π (or 180°). We will revisit the periodic nature of trigonometric functions later in this book when we discuss their graphs in Chapters 9, 10, and 11.

Here is an example that uses periodic identities.

EXAMPLE

Use periodic identities to find the exact value of the trigonometric expression $\dfrac{\sin(1020°)}{\tan(930°)}$.

> We proceed as follows:

$$\frac{\sin(1020°)}{\tan(930°)} = \frac{\sin(300° + 2 \cdot 360°)}{\tan(30° + 5 \cdot 180°)} = \frac{\sin(300°)}{\tan(30°)} = \frac{-\sqrt{3}/2}{\sqrt{3}/3} = -\frac{3}{2}$$

The **even-odd identities** are the following:

$$\sin(-\theta) = -\sin\theta \qquad \cos(-\theta) = \cos\theta$$
$$\tan(-\theta) = -\tan\theta$$

BTW

The even-odd identities also are known as the negative angle identities.

Let's try an example that involves using even-odd identities.

EXAMPLE

> Use even-odd identities to determine the exact value of the trigonometric expression $\sin\left(-\dfrac{\pi}{3}\right) + \cos\left(-\dfrac{\pi}{4}\right)$.

> We proceed as follows:

$$\sin\left(-\frac{\pi}{3}\right) + \cos\left(-\frac{\pi}{4}\right) = -\sin\left(\frac{\pi}{3}\right) + \cos\left(\frac{\pi}{4}\right) = -\frac{\sqrt{3}}{2} + \frac{\sqrt{2}}{2} = \frac{\sqrt{2} - \sqrt{3}}{2}$$

The Pythagorean Identities

The following **Pythagorean identities** can be obtained from the trigonometric ratios discussed in Chapter 6.

$$\sin^2\theta + \cos^2\theta = 1 \qquad \tan^2\theta + 1 = \sec^2\theta \qquad \cot^2\theta + 1 = \csc^2\theta$$

We can derive the identity $\sin^2\theta + \cos^2\theta = 1$, as shown next.

Consider the following figure.

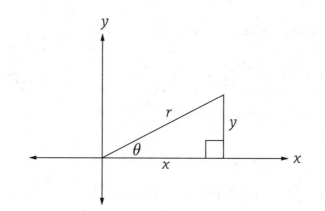

From the definitions of the trigonometric ratios, $\sin\theta = \dfrac{y}{r}$ and

$\cos\theta = \dfrac{x}{r}$. It follows that $\sin^2\theta + \cos^2\theta = \left(\dfrac{y}{r}\right)^2 + \left(\dfrac{x}{r}\right)^2 = \dfrac{y^2 + x^2}{r^2}$. By the

Pythagorean theorem, $y^2 + x^2 = r^2$. Thus, $\sin^2\theta + \cos^2\theta = \dfrac{r^2}{r^2} = 1$.

IRL In calculus, substitution of Pythagorean identities is used to eliminate radicals from certain algebraic expressions. For example, in the radical

$\sqrt{a^2 - x^2}$, we can substitute $x = a\sin\theta$ with $a > 0$ and $-\dfrac{\pi}{2} \le \theta \le \dfrac{\pi}{2}$ to obtain

$\sqrt{a^2 - x^2} = \sqrt{a^2 - (a\sin\theta)^2} = \sqrt{a^2 - a^2\sin^2\theta} = \sqrt{a^2(1 - \sin^2\theta)} =$

$\sqrt{a^2(\cos^2\theta)} = a\cos\theta.$

In the following example, we use the Pythagorean identity $\sin^2\theta + \cos^2\theta = 1$ to verify the Pythagorean identity $\tan^2\theta + 1 = \sec^2\theta$.

EXAMPLE

▶ Verify that $\tan^2\theta + 1 = \sec^2\theta$ is an identity.

▶ Starting with the left side, we proceed as follows:

$$\tan^2\theta + 1 = \left(\frac{\sin\theta}{\cos\theta}\right)^2 + 1 = \frac{\sin^2\theta}{\cos^2\theta} + 1$$

$$= \frac{\sin^2\theta + \cos^2\theta}{\cos^2\theta} = \frac{1}{\cos^2\theta}$$

$$= \sec^2\theta$$

▶ Hence, $\tan^2\theta + 1 = \sec^2\theta$ is an identity.

Sum and Difference Formulas for the Sine Function

The following are the sum and difference formulas for the sine function:

$$\sin(\theta + \varphi) = \sin\theta\cos\varphi + \cos\theta\sin\varphi$$
$$\sin(\theta - \varphi) = \sin\theta\cos\varphi - \cos\theta\sin\varphi$$

Here is an example where the angle is expressed in degrees.

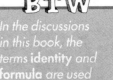

BTW

In the discussions in this book, the terms **identity** and **formula** are used interchangeably.

EXAMPLE

▶ Find the exact value of $\sin 15°$.

▶ First, we express $15°$ as the difference of two special angles: $15° = (45° - 30°)$. Then, we have $\sin 15° = \sin(45° - 30°)$.

BTW

We think in terms of special angles when a question requires us to find an "exact value."

▶ Next, we apply the difference formula for the sine function to obtain

$$\sin 15° = \sin(45° - 30°) = \sin 45°\cos 30° - \cos 45°\sin 30°$$

$$= \frac{\sqrt{2}}{2}\cdot\frac{\sqrt{3}}{2} - \frac{\sqrt{2}}{2}\cdot\frac{1}{2} = \frac{\sqrt{6} - \sqrt{2}}{4}.$$

Let's try an example in which the angle is expressed in radians.

EXAMPLE

▶ Find the exact value of $\sin\dfrac{\pi}{12}$.

▶ First, we express $\dfrac{\pi}{12}$ as the difference of two special angles:

$\dfrac{\pi}{12} = \left(\dfrac{\pi}{3} - \dfrac{\pi}{4}\right)$. Then, we have $\sin\dfrac{\pi}{12} = \sin\left(\dfrac{\pi}{3} - \dfrac{\pi}{4}\right)$.

▶ Next, we apply the difference formula for the

sine function to obtain $\sin\dfrac{\pi}{12} = \sin\left(\dfrac{\pi}{3} - \dfrac{\pi}{4}\right) =$

$\sin\dfrac{\pi}{3}\cos\dfrac{\pi}{4} - \cos\dfrac{\pi}{3}\sin\dfrac{\pi}{4} = \dfrac{\sqrt{3}}{2}\cdot\dfrac{\sqrt{2}}{2} - \dfrac{1}{2}\cdot\dfrac{\sqrt{2}}{2} = \dfrac{\sqrt{6} - \sqrt{2}}{4}.$

The following example shows a clever use of the sum formula.

EXAMPLE

▶ Write $\sin 5\theta\cos 3\theta + \cos 5\theta\sin 3\theta$ as a single function of $k\theta$ for some integer k.

▶ We recognize $\sin 5\theta\cos 3\theta + \cos 5\theta\sin 3\theta$ as an application of the sine of the sum of two angles formula.

▶ Thus, $\sin 5\theta\cos 3\theta + \cos 5\theta\sin 3\theta = \sin(5\theta + 3\theta) = \sin 8\theta$.

The following example shows a way to reduce a trigonometric expression to a simpler one.

EXAMPLE

▶ Use a sum or difference formula to verify the identity: $\sin(180° - \theta) = \sin\theta$.

▶ Starting with the left side, we proceed as follows: $\sin(180° - \theta) =$
$\sin 180°\cos\theta - \cos 180°\sin\theta = 0\cdot\cos\theta - (-1)\sin\theta = \sin\theta$.

▶ Thus, $\sin(180° - \theta) = \sin\theta$ is an identity.

Sum and Difference Formulas for the Cosine Function

The following are the sum and difference formulas for the cosine function:

$$\begin{aligned} \cos(\theta + \phi) &= \cos\theta\cos\phi - \sin\theta\sin\phi \\ \cos(\theta - \phi) &= \cos\theta\cos\phi + \sin\theta\sin\phi \end{aligned}$$

Here is an example where the angle is expressed in degrees.

EXAMPLE

▶ Find the exact value of $\cos 75°$.

▶ First, we express $75°$ as the sum of two special angles: $75° = (45° + 30°)$. Then, we have $\cos 75° = \cos(45° + 30°)$.

▶ Next, we apply the sum formula for the cosine function to obtain
$$\cos 75° = \cos(45° + 30°) = \cos 45° \cos 30° - \sin 45° \sin 30° =$$
$$\frac{\sqrt{2}}{2} \cdot \frac{\sqrt{3}}{2} - \frac{\sqrt{2}}{2} \cdot \frac{1}{2} = \frac{\sqrt{6} - \sqrt{2}}{4}.$$

Here is an example where the angle is expressed in radians.

EXAMPLE

▶ Find the exact value of $\cos\dfrac{\pi}{12}$.

▶ First, we express $\dfrac{\pi}{12}$ as the difference of two special angles: $\dfrac{\pi}{12} = \left(\dfrac{\pi}{4} - \dfrac{\pi}{6}\right)$. Then, we have $\cos\dfrac{\pi}{12} = \cos\left(\dfrac{\pi}{4} - \dfrac{\pi}{6}\right)$.

▶ Next, we apply the difference formula for the cosine function to obtain
$$\cos\frac{\pi}{12} = \cos\left(\frac{\pi}{4} - \frac{\pi}{6}\right) = \cos\frac{\pi}{4}\cos\frac{\pi}{6} + \sin\frac{\pi}{4}\sin\frac{\pi}{6} =$$
$$\frac{\sqrt{2}}{2} \cdot \frac{\sqrt{3}}{2} + \frac{\sqrt{2}}{2} \cdot \frac{1}{2} = \frac{\sqrt{6} + \sqrt{2}}{4}.$$

In the following example, we work backward, so to speak.

EXAMPLE

▶ Write $\cos 2\theta \cos 3\theta - \sin 2\theta \sin 3\theta$ as a single function of $k\theta$ for some integer k.

▶ We recognize $\cos 2\theta \cos 3\theta - \sin 2\theta \sin 3\theta$ as an application of the cosine of the sum of two angles formula.

▶ Thus, $\cos 2\theta \cos 3\theta - \sin 2\theta \sin 3\theta = \cos(2\theta + 3\theta) = \cos 5\theta$.

Sum and Difference Formulas for the Tangent Function

We can derive the sum and difference formulas for the tangent function from those for the sine and cosine functions. The derivation of $\tan(\theta + \varphi)$ follows as an example of using established identities to expand our inventory of useful identities.

$$
\begin{aligned}
\tan(\theta + \varphi) &= \frac{\sin(\theta + \varphi)}{\cos(\theta + \varphi)} \\[2mm]
&= \frac{\sin \theta \cos \varphi + \cos \theta \sin \varphi}{\cos \theta \cos \varphi - \sin \theta \sin \varphi} \\[2mm]
&= \frac{\cos \theta \cos \varphi \left(\dfrac{\sin \theta}{\cos \theta} + \dfrac{\sin \varphi}{\cos \varphi} \right)}{\cos \theta \cos \varphi \left(1 - \dfrac{\sin \theta \sin \varphi}{\cos \theta \cos \varphi} \right)} \\[2mm]
&= \frac{\dfrac{\sin \theta}{\cos \theta} + \dfrac{\sin \varphi}{\cos \varphi}}{1 - \dfrac{\sin \theta}{\cos \theta} \cdot \dfrac{\sin \varphi}{\cos \varphi}} \\[2mm]
&= \frac{\tan \theta + \tan \varphi}{1 - \tan \theta \tan \varphi}
\end{aligned}
$$

Thus, we have the following formula:

$$\tan(\theta + \varphi) = \frac{\tan\theta + \tan\varphi}{1 - \tan\theta\tan\varphi}$$

And by a similar manipulation:

$$\tan(\theta - \varphi) = \frac{\tan\theta - \tan\varphi}{1 + \tan\theta\tan\varphi}$$

Let's try an example where the angle is expressed in degrees.

EXAMPLE

▶ Find the exact value of $\tan 15°$.

▶ First, we express $15°$ as the difference of two special angles: $15° = (45° - 30°)$. Then, we have $\tan 15° = \tan(45° - 30°)$.

▶ Next, we apply the difference formula for the tangent function to obtain

$$\tan 15° = \tan(45° - 30°) = \frac{\tan 45° - \tan 30°}{1 + \tan 45° \tan 30°}$$

$$= \frac{1 - \dfrac{1}{\sqrt{3}}}{1 + 1 \cdot \dfrac{1}{\sqrt{3}}} = \frac{\sqrt{3} - 1}{\sqrt{3} + 1}$$

In the next example, we'll use the formula for the tangent of the difference between two angles to verify an equation.

EXAMPLE

▶ Verify that $\dfrac{\tan 65° - \tan 20°}{1 + \tan 65° \tan 20°} = 1.$

▶ We recognize $\dfrac{\tan 65° - \tan 20°}{1 + \tan 65° \tan 20°}$ as an application of the tangent of the difference of two angles formula. Thus, starting with the left side, we proceed as follows:

$$\frac{\tan 65° - \tan 20°}{1 + \tan 65° \tan 20°} = \tan(65° - 20°) = \tan 45° = 1$$

▶ Thus, $\dfrac{\tan 65° - \tan 20°}{1 + \tan 65° \tan 20°} = 1.$

Let's try an example similar to the one we just did, but with an arbitrary angle θ.

EXAMPLE

▶ Write the expression $\dfrac{\tan 2\theta + \tan \theta}{1 - \tan 2\theta \tan \theta}$ as a single function of $k\theta$, where k is an integer.

▶ We recognize $\dfrac{\tan 2\theta + \tan \theta}{1 - \tan 2\theta \tan \theta}$ as an application of the tangent of the sum of two angles formula.

▶ Thus, $\dfrac{\tan 2\theta + \tan \theta}{1 - \tan 2\theta \tan \theta} = \tan(2\theta + \theta) = \tan 3\theta.$

Double-Angle Identities

Using the sum formulas for sine, cosine, and the tangent functions, we can quickly derive the following identities:

$$\sin 2\theta = 2\sin\theta\cos\theta$$

$$\cos 2\theta = \cos^2\theta - \sin^2\theta = 1 - 2\sin^2\theta = 2\cos^2\theta - 1$$

$$\tan 2\theta = \frac{2\tan\theta}{1 - \tan^2\theta}$$

These identities can be used to convert an expression from double angles to single angles or from multiple expressions to a single expression and vice versa.

Look at this example.

EXAMPLE

▶ Express $\dfrac{\tan 3\theta}{1 - \tan^2 3\theta}$ as a single function of $k\theta$, where k is an integer.

▶ We recognize $\dfrac{\tan 3\theta}{1 - \tan^2 3\theta}$ as similar to an application of the double-angle identity for tangent.

▶ Thus, $\dfrac{\tan 3\theta}{1 - \tan^2 3\theta} = \dfrac{1}{2}\left(\dfrac{2\tan 3\theta}{1 - \tan^2 3\theta}\right) = \dfrac{1}{2}\tan 6\theta.$

In the following example, we use a double-angle identity to verify an identity.

EXAMPLE

▶ Verify that $\sin 2\theta = \dfrac{2\tan\theta}{1 + \tan^2\theta}$ is an identity.

▶ Starting with the left side, we proceed as follows:

$$\sin 2\theta = 2\sin\theta\cos\theta = \cos^2\theta\left(2\frac{\sin\theta}{\cos\theta}\right) = \cos^2\theta(2\tan\theta)$$

$$= \frac{1}{\sec^2\theta}(2\tan\theta) = \frac{2\tan\theta}{1 + \tan^2\theta}$$

▶ Thus, $\sin 2\theta = \dfrac{2\tan\theta}{1 + \tan^2\theta}$ is an identity.

Half-Angle Identities

As shown next, we can rearrange the double-angle cosine formulas, $\cos 2\theta = 2\cos^2\theta - 1$ and $\cos 2\theta = 1 - 2\sin^2\theta$, to develop half-angle identities.

First, transform the two equations as follows:

$$\cos^2\theta = \frac{1 + \cos 2\theta}{2}$$

$$\sin^2\theta = \frac{1 - \cos 2\theta}{2}$$

Next, substitute $\dfrac{\theta}{2}$ for θ into each of the equations to obtain

$$\cos^2\frac{\theta}{2} = \frac{1 + \cos\theta}{2}$$

$$\sin^2\frac{\theta}{2} = \frac{1 - \cos\theta}{2}$$

Then, for each of the two equations, take the square root of both sides, yielding the following two half-angle identities:

$$\cos\frac{\theta}{2} = \pm\sqrt{\frac{1 + \cos\theta}{2}}$$

$$\sin\frac{\theta}{2} = \pm\sqrt{\frac{1 - \cos\theta}{2}}$$

Be mindful that taking the square roots introduces algebraic signs that are dependent on the quadrant of the involved angle.

IRL The Mach number of an aircraft, *M*, is the ratio of the speed of the aircraft to the speed of sound at that altitude and temperature. *Mach 1* indicates the speed of sound, *Mach 2* indicates twice the speed of sound, and so forth. When an aircraft travels faster than the speed of sound (*M* > 1), it creates a shock wave that forms a moving cone of pressurized air molecules around the aircraft that move outward and rearward in all directions and extend all the way to the ground. On the ground, a sonic boom is heard at the intersection of the full width of the base of the cone with the ground. If the vertex angle of the cone is θ, then $\sin^2 \dfrac{\theta}{2} = \dfrac{1}{M}$.

During the 1950s and into the 1960s, sonic booms caused by Air Force supersonic jets were a frequent occurrence across the nation, but a public outcry protesting the noise forced the passage of The Noise Control Act of 1972.

In addition, $\tan \dfrac{\theta}{2} = \dfrac{\sin\theta}{1+\cos\theta} = \dfrac{1-\cos\theta}{\sin\theta}$, which can be derived from the two formulas provided earlier.

Here is an example using a half-angle identity.

EXAMPLE

▶ Use a half-angle identity to find the exact value of $\cos 105°$.

▶ First, we express $105°$ as one-half of a special angle: $105° = \dfrac{210°}{2}$. Then, we have $\cos 105° = \cos\left(\dfrac{210°}{2}\right)$. Next, we apply the half-angle formula for cosine.

▶ Because the cosine is negative in quadrant II, we have

$$\cos 105° = \cos\left(\frac{210°}{2}\right) = -\sqrt{\frac{1+\cos 210°}{2}} = -\sqrt{\frac{1+\left(-\dfrac{\sqrt{3}}{2}\right)}{2}}$$

$$= -\sqrt{\frac{1-\dfrac{\sqrt{3}}{2}}{2}} = -\sqrt{\frac{2-\sqrt{3}}{4}} = -\frac{1}{2}\sqrt{2-\sqrt{3}}.$$

Look at the application of a half-angle formula in the following example.

EXAMPLE

If $\sin\dfrac{\theta}{2} = \dfrac{3}{\sqrt{10}}$ and $\cos\theta$ is negative, find $\tan\theta$ for $0 \le \theta \le 360°$.

Because $\cos\theta$ is negative and $\sin\dfrac{\theta}{2}$ is positive, we know θ is in

quadrant II. Thus, $\sin\dfrac{\theta}{2} = \dfrac{3}{\sqrt{10}} = \sqrt{\dfrac{1-\cos\theta}{2}}$. Squaring across the

board yields $\sin^2\dfrac{\theta}{2} = \dfrac{9}{10} = \dfrac{1-\cos\theta}{2}$.

Next, we solve $\dfrac{9}{10} = \dfrac{1-\cos\theta}{2}$ for $\cos\theta$ to obtain $\cos\theta = -\dfrac{4}{5}$. Using

techniques from Chapter 6, it follows that $\tan\theta = -\dfrac{3}{4}$.

The following is another example of verifying an identity.

EXAMPLE

Verify that $\tan\dfrac{\theta}{2}\sin\theta = \dfrac{\tan\theta - \sin\theta}{\sin\theta\sec\theta}$ is an identity.

Starting with the right side and converting all trigonometric functions to sines and cosines, we proceed as follows:

$$\dfrac{\tan\theta - \sin\theta}{\sin\theta\sec\theta} = \dfrac{\dfrac{\sin\theta}{\cos\theta} - \sin\theta}{\sin\theta\dfrac{1}{\cos\theta}} = \dfrac{\sin\theta\left(\dfrac{1}{\cos\theta} - 1\right)}{\sin\theta\dfrac{1}{\cos\theta}} = \dfrac{\dfrac{1}{\cos\theta} - 1}{\dfrac{1}{\cos\theta}}$$

$$= \dfrac{\dfrac{1-\cos\theta}{\cos\theta}}{\dfrac{1}{\cos\theta}} = \dfrac{1-\cos\theta}{1} = \dfrac{\sin\theta(1-\cos\theta)}{\sin\theta \cdot 1} = \sin\theta\left(\dfrac{1-\cos\theta}{\sin\theta}\right)$$

$$= \left(\dfrac{1-\cos\theta}{\sin\theta}\right)\sin\theta = \tan\dfrac{\theta}{2}\sin\theta$$

Thus, $\tan\dfrac{\theta}{2}\sin\theta = \dfrac{\tan\theta - \sin\theta}{\sin\theta\sec\theta}$ is an identity.

EXERCISES

EXERCISE 7-1

Fill in each blank to make a true statement.

1. A trigonometric identity is an equation that is true for all values of the _____ in the domains of the associated trigonometric functions.

2. An underlying assumption of a trigonometric identity is that no expression in any denominator is equal to _____.

3. We verify an identity by transforming one side of the identity until it is _____ to the other side.

4. It takes only one _____ to establish that a statement is not an identity.

5. Verifying an identity _____ (can, cannot) be accomplished by repeatedly substituting in domain values to obtain a true equation.

EXERCISE 7-2

For questions 1 to 6, use identities to express each of the following in terms of one trigonometric function.

1. $\dfrac{\sin^3\theta}{\cos^3\theta}$

2. $\dfrac{\sec\theta}{\tan\theta}$

3. $\sec\theta\cot\theta$

4. $\csc\theta\tan\theta$

5. $\dfrac{\csc\theta}{\cot\theta}$

6. $\dfrac{\cot^2\theta}{\csc^2\theta}$

For questions 7 to 10, verify that the statement is an identity.

7. $\dfrac{\sin^2 2\theta \cot 2\theta}{\cos 2\theta} = \sin 2\theta$

8. $(\sin\theta)(\cot^2\theta)(\sec^2\theta) = \csc\theta$

9. $\dfrac{\sec\theta}{\cot\theta} = \dfrac{\sin\theta}{\cos^2\theta}$

10. $\dfrac{\cot^2\theta}{\sec^2\theta} = \dfrac{\cos^4\theta}{\sin^2\theta}$

EXERCISE 7-3

For questions 1 to 5, determine the acute angle θ using cofunction identities.

1. $\cos\theta = \sin 41°$

2. $\tan\theta = \cot\dfrac{7\pi}{30}$

3. $\cos 75° = \sin\theta$

4. $\sec\theta = \csc\dfrac{2\pi}{5}$

5. $\cot\dfrac{16\pi}{45} = \tan\theta$

For questions 6 to 10, use periodic identities to determine the exact value of the trigonometric expression.

6. $\sin 390°$

7. $\sec\dfrac{9\pi}{4}$

8. $\tan 1320°$

9. $\dfrac{\sqrt{2}\cot\dfrac{9\pi}{4}}{\sin\dfrac{15\pi}{4}}$

10. $\cos^2 2190°$

For questions 11 to 15, use even-odd identities to determine the exact value of the trigonometric expression.

11. $\sin(-45°) - \cos(-60°)$

12. $\dfrac{\csc\left(-\dfrac{\pi}{3}\right)}{\sin\left(-\dfrac{\pi}{2}\right)}\cos\left(-\dfrac{\pi}{6}\right)$

13. $\sqrt{3}\tan\left(-\dfrac{\pi}{3}\right)$

14. $4\sin(90°)\left[\csc(-90°) - \sec(-180°)\right]$

15. $-\sqrt{2}\sec(-45°)\sin(-30°)$

EXERCISE 7-4

For questions 1 to 10, verify the identity.

1. $\cot^2\theta + 1 = \csc^2\theta$

2. $\csc\theta + \sin\theta = \dfrac{\sin^2\theta + 1}{\sin\theta}$

3. $1 + \tan^2\theta = \dfrac{1}{\cos^2\theta}$

4. $\cos^4\theta - \sin^4\theta = \sin^2\theta - \cos^2\theta$

5. $\cos^2\theta(1 + \tan^2\theta) = 1$

6. $\dfrac{\sin^2\theta}{1 - \cos\theta} = \dfrac{1 + \sec\theta}{\sec\theta}$

7. $\dfrac{1}{1 - \sin\theta} + \dfrac{1}{1 + \sin\theta} = \dfrac{2}{\cos^2\theta}$

8. $(\csc^2\theta - 1)(\sec^2\theta - 1) = 1$

9. $\dfrac{1 + \tan^2\theta}{\csc^2\theta} = \tan^2\theta$

10. $\dfrac{\sin^2\theta}{\tan^2\theta - \sin^2\theta} = \cot^2\theta$

EXERCISE 7-5

For questions 1 to 5, use a sum or difference formula for the sine of two angles to find the exact value of the trigonometric function.

1. $\sin 105°$

2. $\sin 75°$

3. $\sin \dfrac{5\pi}{12}$

4. $\sin \dfrac{13\pi}{12}$

5. $\sin 120°$

For questions 6 to 10, write the expression as one trigonometric function of $k\theta$ for some integer k.

6. $\sin 5\theta \cos\theta + \cos 5\theta \sin\theta$

7. $\sin 10\theta \cos 6\theta - \cos 10\theta \sin 6\theta$

8. $\sin\dfrac{3}{4}\theta \cos\dfrac{5}{4}\theta + \cos\dfrac{3}{4}\theta \sin\dfrac{5}{4}\theta$

9. $\sin 5\theta \cos 4\theta - \cos 5\theta \sin 4\theta$

10. $\sin\dfrac{2}{3}\theta \cos\dfrac{1}{3}\theta + \cos\dfrac{2}{3}\theta \sin\dfrac{1}{3}\theta$

For questions 11 to 15, verify the identity.

11. $\sin(180° + \theta) = -\sin\theta$

12. $\sin(360° - \theta) = -\sin\theta$

13. $\sin(90° - \theta) = \cos\theta$

14. $\sin\left(\dfrac{3\pi}{2} + \theta\right) = -\cos\theta$

15. $\sin\left(\theta - \dfrac{\pi}{6}\right) = \dfrac{\sqrt{3}\sin\theta - \cos\theta}{2}$

EXERCISE 7-6

For questions 1 to 5, use a sum or difference formula for the cosine of two angles to find the exact value of the trigonometric function.

1. $\cos 195°$

2. $\cos\dfrac{5\pi}{12}$

3. $\cos 165°$

4. $\cos\dfrac{\pi}{12}$

5. $\cos 120°$

For questions 6 to 10, write the expression as one trigonometric function of $k\theta$ for some integer k.

6. $\cos\theta\cos 2\theta + \sin\theta\sin 2\theta$

7. $\cos\dfrac{3}{4}\theta\cos\dfrac{5}{4}\theta - \sin\dfrac{3}{4}\theta\sin\dfrac{5}{4}\theta$

8. $\cos\pi\theta\cos(\pi - 4)\theta + \sin\pi\theta\sin(\pi - 4)\theta$

9. $\cos 8\theta\cos 6\theta + \sin 8\theta\sin 6\theta$

10. $\cos(\theta + 70°)\cos(\theta - 70°) - \sin(\theta + 70°)\sin(\theta - 70°)$

For questions 11 to 15, verify the identity.

11. $\cos(180° - \theta) = -\cos\theta$

12. $\cos(180° + \theta) = -\cos\theta$

13. $\cos(360° - \theta) = \cos\theta$

14. $\cos\left(\dfrac{5\pi}{2} + \theta\right) = -\sin\theta$

15. $\cos\left(\theta - \dfrac{\pi}{3}\right) = \dfrac{\cos\theta + \sqrt{3}\sin\theta}{2}$

EXERCISE 7-7

For questions 1 to 5, use a sum or difference formula for the tangent of two angles to find the exact value of the trigonometric function.

1. $\tan 195°$

2. $\tan 75°$

3. $\tan 105°$

4. $\tan\dfrac{\pi}{12}$

5. $\tan 165°$

For questions 6 to 10, verify that the statement is correct.

6. $\dfrac{\tan 130° + \tan 50°}{1 - \tan 130° \tan 50°} = 0$

7. $\dfrac{\tan 110° - \tan 50°}{1 + \tan 110° \tan 50°} = \sqrt{3}$

8. $\dfrac{\tan 115° - \tan 70°}{1 + \tan 115° \tan 70°} = 1$

9. $\dfrac{\tan 100° + \tan 50°}{1 - \tan 100° \tan 50°} = -\dfrac{1}{\sqrt{3}}$

10. $\dfrac{\tan 95° + \tan 40°}{1 - \tan 95° \tan 40°} = -1$

For questions 11 to 15, write the expression as one trigonometric function of $k\theta$, where k is an integer.

11. $\dfrac{\tan 4\theta + \tan 5\theta}{1 - \tan 4\theta \tan 5\theta}$

12. $\dfrac{\tan 5\theta - \tan \theta}{1 + \tan 5\theta \tan \theta}$

13. $\dfrac{\tan 8\theta + \tan 6\theta}{1 - \tan 8\theta \tan 6\theta}$

14. $\dfrac{\tan \dfrac{4}{3}\theta - \tan \dfrac{1}{3}\theta}{1 + \tan \dfrac{4}{3}\theta \tan \dfrac{1}{3}\theta}$

15. $\dfrac{\tan \dfrac{5}{4}\theta + \tan \dfrac{3}{4}\theta}{1 - \tan \dfrac{5}{4}\theta \tan \dfrac{3}{4}\theta}$

EXERCISE 7-8

For questions 1 to 5, write the given expression as one trigonometric function of $k\theta$, where k is an integer.

1. $\dfrac{4\tan 2\theta}{1 - \tan^2 2\theta}$

2. $\cos^2 5\theta - \sin^2 5\theta$

3. $2\sin 2\theta \cos 2\theta$

4. $\dfrac{2\tan \theta}{1 - \tan^2 \theta}$

5. $1 - 2\sin^2 3\theta$

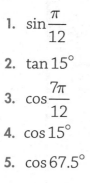

For questions 6 to 10, verify the identity.

6. $\sin 2\theta \csc \theta = 2\cos \theta$

7. $\sec^2 \theta = \dfrac{2}{1 + \cos 2\theta}$

8. $\cot(4\theta) = \dfrac{1 - \tan^2(2\theta)}{2\tan(2\theta)}$

9. $\csc 2\theta = \dfrac{1}{2}\cot \theta + \dfrac{1}{2}\tan \theta$

10. $\sin 4\theta = 4\cos^3 \theta \sin \theta - 4\cos \theta \sin^3 \theta$

EXERCISE 7-9

Use a half-angle identity to find the exact value of each trigonometric function.

1. $\sin \dfrac{\pi}{12}$

2. $\tan 15°$

3. $\cos \dfrac{7\pi}{12}$

4. $\cos 15°$

5. $\cos 67.5°$

Flashcard App

8 Trigonometric Functions of Real Numbers

MUST KNOW

⚡ We define a trigonometric function of a real number so that its value at a real number *t* equals its trigonometric function value at an angle of *t* radians (provided that value exists).

⚡ Trigonometric function definitions and identities are valid whether the inputs are angles or real numbers.

⚡ The domain of a trigonometric function of a real number is the set of all real numbers for which the function is defined.

⚡ The trigonometric functions are periodic; that is, they repeat their values in regular intervals or periods.

n previous lessons, the input of a trigonometric function was an angle, often denoted by θ and measured in either degrees or radians. In this chapter we move away from using angles as the inputs of trigonometric functions to using real numbers as inputs. Conceiving of the trigonometric functions as functions of real numbers is a major step in the study of trigonometry.

Definitions and Basic Concepts of Trigonometric Functions of Real Numbers

To define a trigonometric function for which the input is a real number t, let the value of the trigonometric function be equal to its trigonometric function value at an angle of t radians, provided that value exists.

Using this approach, we have the fortunate outcome that the definitions of the trigonometric functions and the identities that have already been established (and the new ones that may come later) remain the same, and they are valid whether the inputs are angles or real numbers.

BTW

Trigonometric functions of real numbers are widely used in calculus and higher mathematics.

Let's try an example in which the input is a real number expressed in terms of π (so that the real number looks deviously like a special angle).

EXAMPLE

▶ What is the exact value of $\sin\dfrac{\pi}{3}$, where $\dfrac{\pi}{3}$ is a real number?

▶ $\sin\dfrac{\pi}{3}$, where $\dfrac{\pi}{3}$ is a real number, equals $\dfrac{\sqrt{3}}{2}$, which is the same value as the sine of $\dfrac{\pi}{3}$ radians.

Here is another example where the input is a real number expressed in terms of π.

EXAMPLE

▶ What is the exact value of $\sec\dfrac{5\pi}{4}$, where $\dfrac{5\pi}{4}$ is a real number?

▶ $\sec\dfrac{5\pi}{4}$, where $\dfrac{5\pi}{4}$ is a real number, equals $-\sqrt{2}$, which is the same value as the secant of $\dfrac{5\pi}{4}$ radians.

When you are evaluating trigonometric functions of real numbers that you don't recognize as corresponding to special angles, set your calculator to radian mode. Then, use the calculator's trigonometric function keys (and other calculator keys, if needed) to evaluate the function. See Appendix A for instructions using the TI-84 Plus calculator.

EXAMPLE

▶ Use a calculator to evaluate $\sin 3.5$, $\cos 3.5$, and $\tan 3.5$. Round answers to three decimal places.

▶ $\sin 3.5 \approx -0.351$, $\cos 3.5 \approx -0.936$, and $\tan 3.5 \approx 0.375$.

EXAMPLE

▶ Use a calculator to evaluate $\sec 3.5$, $\csc 3.5$, and $\cot 3.5$. Round answers to three decimal places.

▶ Most calculators do not have built-in secant, cosecant, and cotangent functions. As shown in Appendix A, for the TI-84 Plus calculator, we use the $\boxed{x^{-1}}$ key with the associated reciprocal identities to evaluate these functions.

▶ Thus, $\sec 3.5 = (\cos 3.5)^{-1} \approx -1.068$, $\csc 3.5 = (\sin 3.5)^{-1} \approx -2.851$, and $\cot 3.5 = (\tan 3.5)^{-1} \approx 2.670$.

The domain of a trigonometric function of a real number is the set of all real numbers for which the function is defined. Thus, for the functions $y = \sin t$ and $y = \cos t$, t can be any real number. For the functions $y = \tan t$ and $y = \sec t$, t can be any real number *except* the real numbers $\frac{n\pi}{2}$, where n is an odd integer. For the functions $y = \cot t$ and $y = \csc t$, t can be any real number *except* the real numbers $n\pi$, where n is any integer.

 **IRL** | Many applications of the trigonometric functions in calculus, engineering, and the sciences require that the inputs be real numbers, rather than angles. For instance, the sine and cosine functions are used in modeling real-life situations such as harmonic motion, electric currents, tides, and weather patterns, in which their inputs are real numbers based on time.

We have now defined trigonometric functions as functions of angles and as functions of real numbers. In your remaining work in trigonometry, be prepared to draw on your understanding of both perspectives.

Periodic Functions

The trigonometric functions are distinctive types of functions because they are periodic. Periodic functions repeat their values in regular intervals or periods. Specifically, a **periodic function** is a nonconstant function f in which there exists a positive constant P such that for all x in the domain of f, $f(x + nP) = f(P)$, where n is an integer. The least number p for which this is true is the **period** of f.

The functions $y = \sin t$, $y = \cos t$, $y = \sec t$, and $y = \csc t$ are periodic functions, each with period 2π. The functions $y = \tan t$ and $y = \cot t$ are periodic functions, each with period π.

The graph of a periodic function displays a repetitive, cyclical pattern that coincides with the period of the function. Each **cycle** is one repetition of the periodic pattern and has a horizontal length equal to the function's period.

The graph of a periodic function that oscillates equally above and below a horizontal line has an **amplitude** that is the maximum height of the graph above the horizontal line. It equals one-half the absolute difference between the function's maximum value and its minimum value; that is, amplitude $= \frac{1}{2}$|maximum value $-$ minimum value|.

The **equation of the midline** about which the graph cycles is the equation of the horizontal line halfway between the maximum and the minimum values; thus, the equation of the midline is,

$$y = \frac{\text{maximum value} + \text{minimum value}}{2}.$$

EXAMPLE

▶ Determine the period, amplitude, and equation of the midline for the graph shown next.

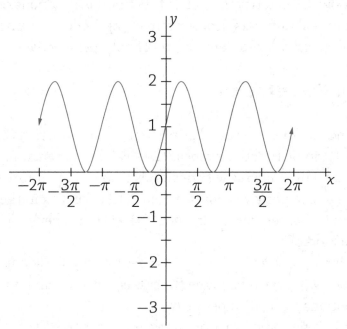

▶ The graph displays a repetitive pattern that repeats every π units, indicating the period is π. The graph has a maximum value of 2 and a minimum value of 0, indicating its amplitude is $\frac{1}{2}|$maximum value $-$ minimum value$| = \frac{1}{2}|2 - 0| = 1$. The midline equation is

$$y = \frac{\text{maximum value} + \text{minimum value}}{2}$$

$$= \frac{2+0}{2} = 1.$$

BTW

Not all periodic functions have an amplitude and a midline. For example, the tangent does not have an amplitude (or a midline) because it does not have a maximum or minimum value. Its amplitude is undefined.

EXERCISES

EXERCISE 8-1

For questions 1 to 5, for the given real number t, find the exact values of sin t, cos t, and tan t. Do not use a calculator.

1. $t = \dfrac{5\pi}{3}$

2. $t = -\dfrac{3\pi}{4}$

3. $t = \dfrac{\pi}{6}$

4. $t = -\dfrac{11\pi}{6}$

5. $t = \pi$

For questions 6 to 10, for the given real number t, find the exact values of sec t, csc t, and cot t. Do not use a calculator.

6. $t = \dfrac{5\pi}{3}$

7. $t = -\dfrac{3\pi}{4}$

8. $t = \dfrac{\pi}{6}$

9. $t = -\dfrac{11\pi}{6}$

10. $t = \dfrac{2\pi}{3}$

For questions 11 to 20, use a calculator to evaluate the function. Round answers to three decimal places.

11. $\sin 1$

12. $\cos 20$

13. $\tan(-1000)$

14. $\sec 3$

15. $\cot 4.4$

16. $\sin 6.1$

17. $\cos(-7.2)$

18. $\tan 15.3$

19. $\sec 0.5$

20. $\cot(-2.6)$

EXERCISE 8-2

For questions 1 to 3, answer the question for the graph of the periodic function shown next.

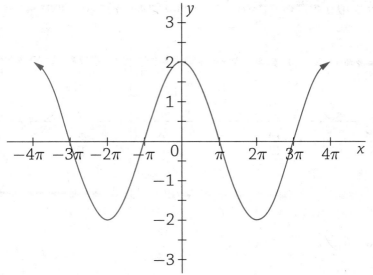

1. What is the function's period?

   a. π

   b. 2π

   c. 3π

   d. 4π

2. What is the function's amplitude?

  a. 1

  b. 2

  c. 4

  d. none

3. What is the equation of the function's midline?

  a. $y = -2$

  b. $y = 0$

  c. $y = 2$

  d. none

Use the following graph of a periodic function to answer questions 4 and 5.

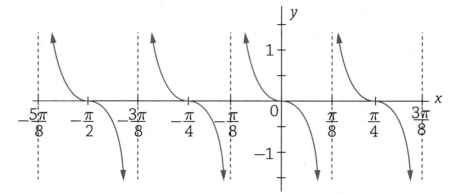

4. What is the function's period?

  a. $\dfrac{\pi}{8}$

  b. $\dfrac{\pi}{4}$

  c. $\dfrac{\pi}{2}$

  d. π

5. What is the function's amplitude?

   a. $\dfrac{1}{2}$

   b. 1

   c. 2

   d. none

 # Graphs of the Sine Function

MUST KNOW

⚡ The graph of $y = \sin x$ provides a visual representation of its properties, including its period, amplitude, maximum value, minimum value, and midline.

⚡ The domain of $y = \sin x$ is $(-\infty, \infty)$, and its range is $[-1, 1]$. Its graph oscillates about the x-axis with a period of 2π, amplitude of 1, and x-intercepts at integer multiples of π.

⚡ Under transformations, the graph of the sine function can be stretched or compressed vertically or horizontally, shifted left or right or up or down, and reflected about a horizontal midline.

⚡ The graph of a sine function provides a model for oscillating, harmonic, or wave motion.

Because we want to graph the trigonometric functions in the x-y plane, from here on out we're going to switch to the traditional symbols x and y when writing the trigonometric functions. Thus, we will write $y = \sin t$ as $y = \sin x$, $y = \cos t$ as $y = \cos x$, $y = \tan t$ as $y = \tan x$, and so forth.

Being able to graph the sine function and knowing its properties, as well as those of its various transformations, are very important skills in trigonometry. In this chapter, we begin by exploring the properties of the graph of the basic sine function $y = \sin x$. Next, we delve into transformations of the basic sine function and the properties of the graphs.

The Graph of $y = \sin x$

To graph the function $y = \sin x$, we create an x-y table using some familiar values of x in the interval from 0 to 2π. We plot the table's ordered pairs and then connect the points with a smooth curve. Hint: To make the graph easier to read, we "stretch" the scale on the y-axis.

x	0	$\dfrac{\pi}{6}$	$\dfrac{\pi}{4}$	$\dfrac{\pi}{2}$	$\dfrac{3\pi}{4}$	$\dfrac{5\pi}{6}$
$\sin x$	0	$\dfrac{1}{2}$	$\dfrac{\sqrt{2}}{2}$	1	$\dfrac{\sqrt{2}}{2}$	$\dfrac{1}{2}$
$(x, \sin x)$	$(0,0)$	$\left(\dfrac{\pi}{6},\dfrac{1}{2}\right)$	$\left(\dfrac{\pi}{4},\dfrac{\sqrt{2}}{2}\right)$	$\left(\dfrac{\pi}{2},1\right)$	$\left(\dfrac{3\pi}{4},\dfrac{\sqrt{2}}{2}\right)$	$\left(\dfrac{5\pi}{6},\dfrac{1}{2}\right)$

(Continued)

x	π	$\dfrac{7\pi}{6}$	$\dfrac{5\pi}{4}$	$\dfrac{3\pi}{2}$	$\dfrac{7\pi}{4}$	$\dfrac{11\pi}{6}$	2π
sin x	0	$-\dfrac{1}{2}$	$-\dfrac{\sqrt{2}}{2}$	-1	$-\dfrac{\sqrt{2}}{2}$	$-\dfrac{1}{2}$	0
(x, sin x)	$(\pi,0)$	$\left(\dfrac{7\pi}{6},-\dfrac{1}{2}\right)$	$\left(\dfrac{5\pi}{4},-\dfrac{\sqrt{2}}{2}\right)$	$\left(\dfrac{3\pi}{2},-1\right)$	$\left(\dfrac{7\pi}{4},-\dfrac{\sqrt{2}}{2}\right)$	$\left(\dfrac{11\pi}{6},-\dfrac{1}{2}\right)$	$(2\pi,0)$

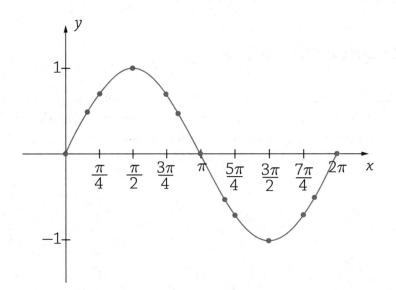

The graph shown represents only a portion of $y = \sin x$. This portion is the **basic cycle** of $y = \sin x$. The entire graph repeats the pattern of the basic cycle, joined end to end in both the positive and negative directions on the x-axis. Following is a graph of $y = \sin x$ that shows several cycles, with the basic cycle plotted with a thicker line style.

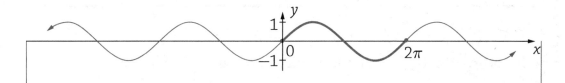

The curve is smooth and continuous with a wavelike appearance. It has no vertical asymptotes, holes, gaps, jumps, or corners.

The domain of $y = \sin x$ is all real numbers. We can easily verify from the following figure that its graph has amplitude of 1 and period of 2π. Its maximum value is 1, and its minimum value is -1. Hence, its range is $-1 \leq y \leq 1$.

> **BTW**
> A vertical **asymptote** is a vertical line that the graph of a function gets closer and closer to in at least one direction along the line.

> **BTW**
> Functions whose graphs have the same wavelike shape of sine functions are **sinusoidal**.

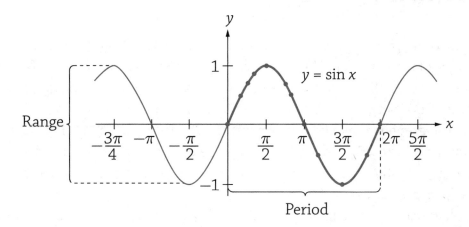

The real zeros of the sine function, meaning the real numbers for which $\sin x = 0$, are integer multiples of π: 0, $\pm\pi$, $\pm2\pi$, $\pm3\pi$, and so on. This information tells us that the graph of $y = \sin x$ has x-intercepts at $\pm n\pi$, where n is an integer.

> **BTW**
> The sine function is odd, meaning $\sin(-x) = -\sin x$. Graphically, an odd function is symmetric about the origin, as exhibited in the graph of $y = \sin x$.

EXAMPLE

▶ For what values of x in the interval $0 \leq x \leq \pi$ is $y = \sin x$ increasing?

▶ In the interval $0 \leq x \leq \pi$, the graph of $y = \sin x$ is increasing

for $0 < x < \dfrac{\pi}{2}$.

IRL First demonstrated by Sir Isaac Newton in 1672, white light is not actually white, but is made up of constituent colors. It is a composition of waves, each of which has a different wavelength and color associated with it. When white light passes through an optical prism, the prism separates the waves according to their wavelengths to form a rainbow. We can represent those light waves graphically using the sine function.

The Graph of $y = A \sin x$

The function $y = A \sin x$ is a transformation of the function $y = \sin x$. The constant A is either a vertical stretch factor or a vertical compression factor, and $|A|$ is the amplitude of the graph of $y = A \sin x$. If $|A| > 1$, the graph is stretched vertically, and if $|A| < 1$, the graph is compressed vertically. If $A < 0$, the graph is reflected over the x-axis. The function $y = A \sin x$ has a maximum value of $|A|$ and a minimum value of $-|A|$. Hence, its range is $-|A| \leq y \leq |A|$. The equation of the midline, the zeros, and the period of $y = A \sin x$ are the same as for $y = \sin x$.

Look at this example in which the function has a vertical stretch factor of 2.

EXAMPLE

▶ What are the period, amplitude, maximum value, minimum value, and midline equation for the function $y = 2 \sin x$?

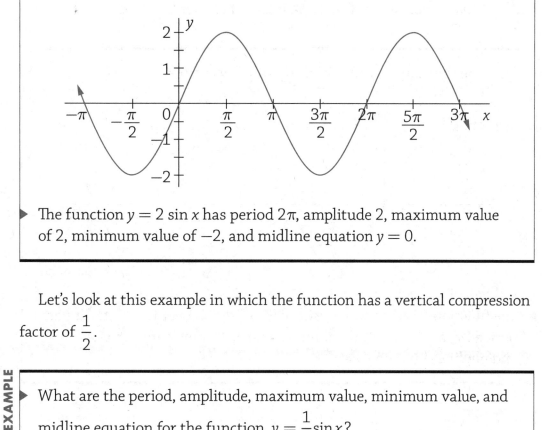

▶ The function $y = 2 \sin x$ has period 2π, amplitude 2, maximum value of 2, minimum value of -2, and midline equation $y = 0$.

Let's look at this example in which the function has a vertical compression factor of $\dfrac{1}{2}$.

EXAMPLE

▶ What are the period, amplitude, maximum value, minimum value, and midline equation for the function $y = \dfrac{1}{2}\sin x$?

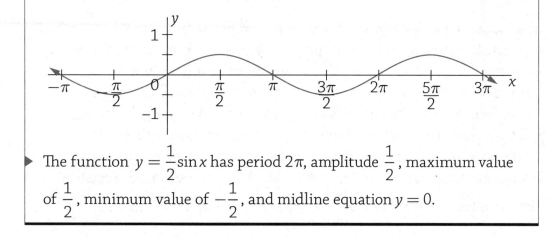

▶ The function $y = \dfrac{1}{2}\sin x$ has period 2π, amplitude $\dfrac{1}{2}$, maximum value of $\dfrac{1}{2}$, minimum value of $-\dfrac{1}{2}$, and midline equation $y = 0$.

Let's try an example where the graph is reflected over the x-axis.

▶ Describe the major features of the graph of $y = \sin(-x)$.

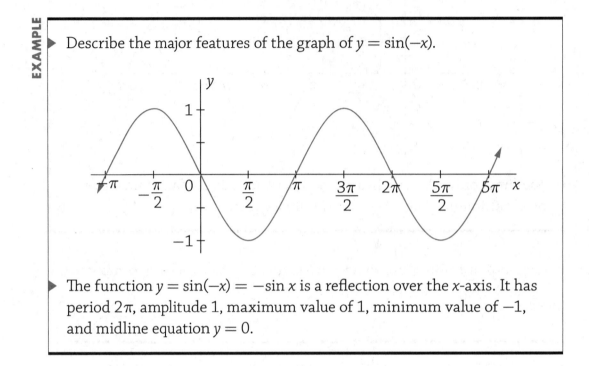

▶ The function $y = \sin(-x) = -\sin x$ is a reflection over the x-axis. It has period 2π, amplitude 1, maximum value of 1, minimum value of -1, and midline equation $y = 0$.

The Graph of $y = A \sin Bx$

The function $y = A \sin Bx$ is a transformation of the function $y = \sin x$. The two constants A and B affect the graph of $y = A \sin Bx$ in different ways. The constant A in $y = A \sin Bx$ has the same impact as it does in $y = A \sin x$ (see the previous lesson in this chapter). It is a vertical stretch (or compression) factor, which affects the graph's amplitude, and if $A < 0$, the graph is a reflection over the x-axis of the corresponding function with $A > 0$.

In contrast, the constant B is either a horizontal stretch factor or a horizontal compression factor, each of which alters the period. Specifically, the period of $y = A \sin Bx$ is $\dfrac{2\pi}{|B|}$. If $|B| > 1$, the graph is compressed horizontally so that it completes a full cycle in a shorter period ($<2\pi$), and if $|B| < 1$, the graph is stretched horizontally so that it completes a full cycle in a longer period ($>2\pi$). If $B < 0$, we can use our knowledge of odd functions to obtain the equivalent function $y = A \sin(-|B|x) = -A \sin(|B|x)$, which is a reflection over the x-axis of $A \sin(|B|x)$.

Look at this example in which the period is compressed horizontally.

▶ What are the period, amplitude, maximum value, minimum value, and midline equation for the function $y = 3 \sin 2x$?

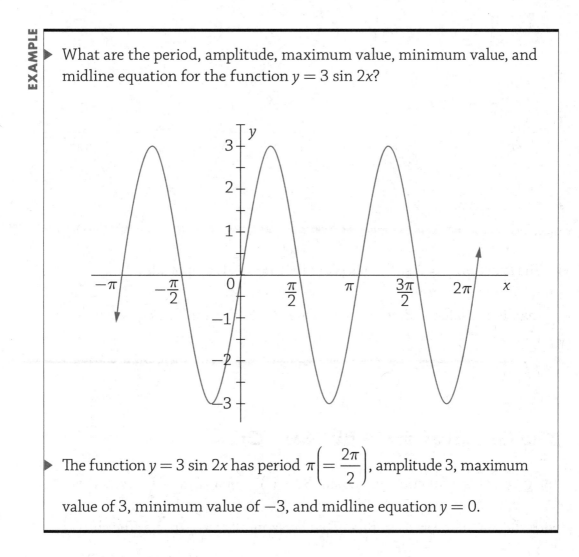

▶ The function $y = 3 \sin 2x$ has period $\pi \left(= \dfrac{2\pi}{2} \right)$, amplitude 3, maximum value of 3, minimum value of -3, and midline equation $y = 0$.

Here's an example in which the period is stretched horizontally.

What are the period, amplitude, maximum value, minimum value, and midline equation for the function $y = 3\sin\dfrac{x}{2}$?

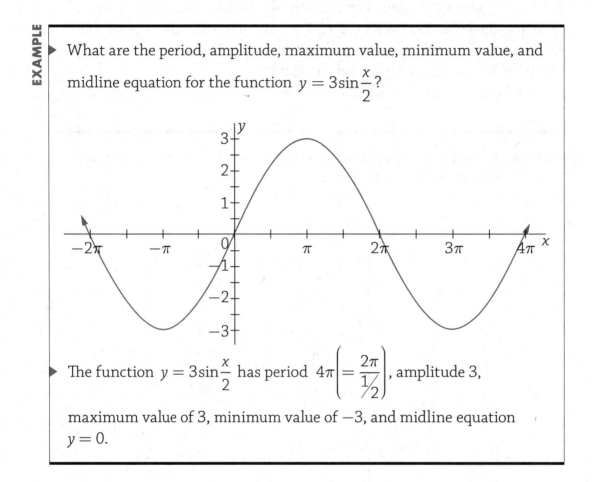

The function $y = 3\sin\dfrac{x}{2}$ has period $4\pi\left(= \dfrac{2\pi}{\frac{1}{2}}\right)$, amplitude 3,

maximum value of 3, minimum value of −3, and midline equation $y = 0$.

The Graph of $y = A\sin(Bx - C)$

The graph of the function $y = A\sin(Bx - C) = A\sin B\left(x - \dfrac{C}{B}\right)$ coincides

with the graph of $y = A\sin Bx$ shifted horizontally by $\dfrac{C}{B}$ units. The shift is to the right when $\dfrac{C}{B} > 0$, and to the left when $\dfrac{C}{B} < 0$. If $A < 0$, the graph is a reflection over the x-axis of the corresponding function with $A > 0$.

The number $\dfrac{C}{B}$ is the **phase shift**. In identifying the phase shift, we find it helpful to write $y = A\sin(Bx - C)$ in **shift form** as $y = A\sin B\left(x - \dfrac{C}{B}\right)$. We solve the equations $x - \dfrac{C}{B} = 0$ and $x - \dfrac{C}{B} = \dfrac{2\pi}{|B|}$ to determine the left and right endpoints, respectively, of an interval that corresponds to one cycle of the graph.

If $B < 0$, we can use the odd function property of the sine function to obtain the equivalent function $y = A\sin\left(-|B|\left(x + \dfrac{C}{|B|}\right)\right) = -A\sin\left(|B|\left(x + \dfrac{C}{|B|}\right)\right)$, which is a reflection over the x-axis of $A\sin\left(|B|\left(x + \dfrac{C}{|B|}\right)\right)$.

In this example, the graph has a shift to the left:

EXAMPLE

▶ Describe the major features of the graph of $y = 3\sin\left(x + \dfrac{\pi}{6}\right)$ and sketch at least one cycle of the function's graph.

▶ The function $y = 3\sin\left(x + \dfrac{\pi}{6}\right) = 3\sin\left(x - \left(-\dfrac{\pi}{6}\right)\right)$ has period 2π, amplitude 3, and a phase shift of $\dfrac{\pi}{6}$ to the left. Solving $x - \left(-\dfrac{\pi}{6}\right) = 0$ and $x - \left(-\dfrac{\pi}{6}\right) = 2\pi$ yields $\left[-\dfrac{\pi}{6}, \dfrac{11\pi}{6}\right]$ as a one-cycle interval of the graph. See the following graph of $y = 3\sin\left(x + \dfrac{\pi}{6}\right) = 3\sin\left(x - \left(-\dfrac{\pi}{6}\right)\right)$.

Here, the portion of the graph corresponding to the interval $\left[-\dfrac{\pi}{6}, \dfrac{11\pi}{6}\right]$ is plotted with the thicker blue line:

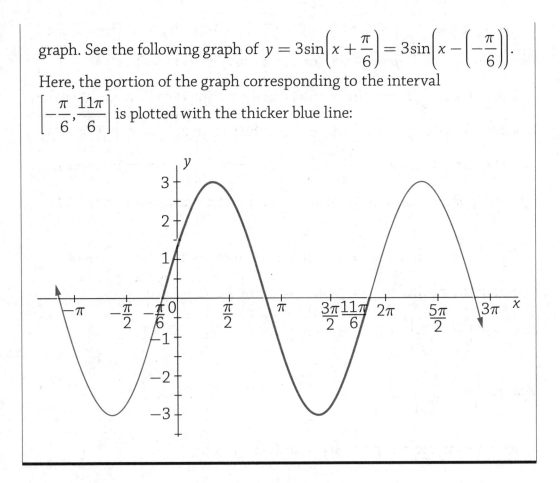

Now let's try this example in which the graph has a shift to the right.

EXAMPLE

▶ Describe the major features of the graph of $y = 2\sin\left(4x - \dfrac{4\pi}{3}\right)$ and sketch at least one cycle of the function's graph.

▶ The function $y = 2\sin\left(4x - \dfrac{4\pi}{3}\right) = 2\sin\left(4\left(x - \dfrac{\pi}{3}\right)\right)$ has period $\dfrac{\pi}{2}\left(= \dfrac{2\pi}{4}\right)$, amplitude 2, and a phase shift of $\dfrac{\pi}{3}$ to the right. Solving $x - \dfrac{\pi}{3} = 0$ and $x - \dfrac{\pi}{3} = \dfrac{\pi}{2}$ yields $\left[\dfrac{\pi}{3}, \dfrac{5\pi}{6}\right]$ as a one-cycle interval of the graph. See the following graph of $y = 2\sin\left(4x - \dfrac{4\pi}{3}\right) = 2\sin\left(4\left(x - \dfrac{\pi}{3}\right)\right)$.

The portion of the graph corresponding to the interval $\left[\dfrac{\pi}{3}, \dfrac{5\pi}{6}\right]$ is plotted with a thick blue line:

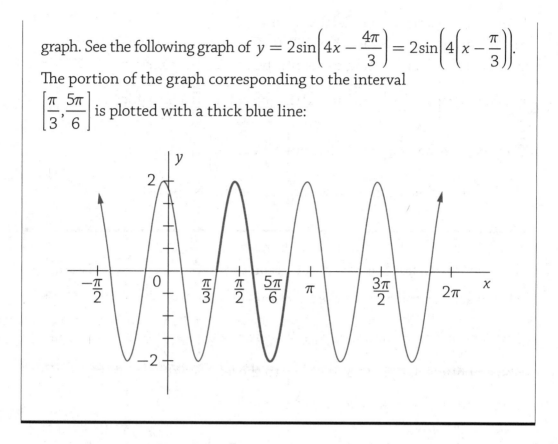

The Graph of $y = A \sin(Bx - C) + D$

The graph of the function $y = A\sin(Bx - C) + D = A\sin B\left(x - \dfrac{C}{B}\right) + D$

coincides with the graph of $y = A\sin(Bx - C)$ shifted vertically by D units. When $D > 0$, the shift is D units upward, and when $D < 0$, the shift is $|D|$ units downward. Instead of $y = 0$, the midline equation for $y = A\sin(Bx - C) + D$ is $y = D$.

If $A < 0$, the graph is a reflection over the line $y = D$ of the corresponding function with $A > 0$. If $B < 0$, we can use the odd function property of the sine function to obtain the equivalent function

$$y = A\sin\left(-|B|\left(x + \frac{C}{|B|}\right)\right) + D = -A\sin\left(|B|\left(x + \frac{C}{|B|}\right)\right) + D,\text{ which is a}$$

reflection over the line $y = D$ of $A\sin\left(|B|\left(x + \frac{C}{|B|}\right)\right) + D$.

Let's look at this example in which the midline is above the x-axis.

EXAMPLE

▶ Describe the major features of the graph of $y = 3\sin\left(x + \frac{\pi}{6}\right) + 2$ and sketch at least one cycle of the function's graph.

▶ The function $y = 3\sin\left(x + \frac{\pi}{6}\right) + 2 = 3\sin\left(x - \left(-\frac{\pi}{6}\right)\right) + 2$ has period 2π, amplitude 3, a phase shift of $\frac{\pi}{6}$ to the left, a vertical shift of 2 upward, and a midline equation of $y = 2$.

▶ Solving $x - \left(-\frac{\pi}{6}\right) = 0$ and $x - \left(-\frac{\pi}{6}\right) = 2\pi$ yields $\left[-\frac{\pi}{6}, \frac{11\pi}{6}\right]$ as a one-cycle interval of the graph. See the following graph of $y = 3\sin\left(x + \frac{\pi}{6}\right) + 2 = 3\sin\left(x - \left(-\frac{\pi}{6}\right)\right) + 2$. We can see that the midline equation $y = 2$ is plotted with a thin blue line, and the portion of the graph corresponding to the interval $\left[-\frac{\pi}{6}, \frac{11\pi}{6}\right]$ is plotted with the thicker blue line:

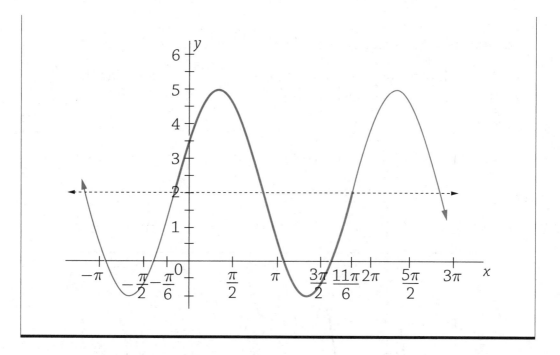

Here we have an example in which the midline is below the x-axis.

▶ Describe the major features of the graph of $y = 2\sin\left(4x - \dfrac{4\pi}{3}\right) - 3$ and sketch at least one cycle of the function's graph.

▶ The function $y = 2\sin\left(4x - \dfrac{4\pi}{3}\right) - 3 = 2\sin\left(4\left(x - \dfrac{\pi}{3}\right)\right) - 3$ has period $\dfrac{\pi}{2}\left(= \dfrac{2\pi}{4}\right)$, amplitude 2, a phase shift of $\dfrac{\pi}{3}$ to the right, a vertical shift of 3 downward, and a midline equation of $y = -3$.

▶ Solving $x - \dfrac{\pi}{3} = 0$ and $x - \dfrac{\pi}{3} = \dfrac{\pi}{2}$ yields $\left[\dfrac{\pi}{3}, \dfrac{5\pi}{6}\right]$ as a one-cycle interval of the graph. See the following graph of $y = 2\sin\left(4x - \dfrac{4\pi}{3}\right) - 3 = 2\sin\left(4\left(x - \dfrac{\pi}{3}\right)\right) - 3$. The midline equation $y = -3$ is plotted with the thin blue line, and the portion of the graph corresponding to the interval $\left[\dfrac{\pi}{3}, \dfrac{5\pi}{6}\right]$ is plotted with a thicker blue:

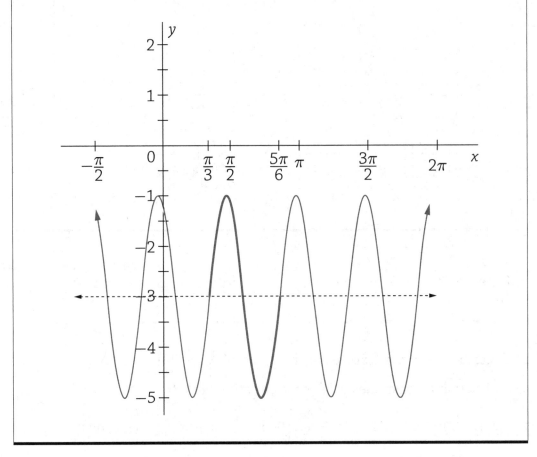

EXERCISES

EXERCISE 9-1

Let's try out our new sin function-graph knowledge.

1. For what values of x in the interval from $0 \leq x \leq 2\pi$ is $\sin x = 0$?

2. For what values of x in the interval from $0 \leq x \leq 2\pi$ is $\sin x = 1$?

3. For what values of x in the interval from $0 \leq x \leq 2\pi$ is $\sin x = -1$?

4. For what values of x in the interval from $0 \leq x \leq 2\pi$ is $y = \sin x$ positive?

5. For what values of x in the interval from $0 \leq x \leq 2\pi$ is $y = \sin x$ negative?

6. For what values of x in the interval from $0 \leq x \leq 2\pi$ is $y = \sin x$ increasing?

7. For what values of x in the interval from $0 \leq x \leq 2\pi$ is $y = \sin x$ decreasing?

8. Is the value -3 in the domain of $y = \sin x$? Yes or No? Justify your answer.

9. Is the value -3 in the range of $y = \sin x$? Yes or No? Justify your answer.

10. Is $\sin(-x) = -\sin x$? Yes or No? Justify your answer.

EXERCISE 9-2

For questions 1 to 5, state the amplitude and range of the given function.

1. $y = 4\sin x$

2. $y = \dfrac{1}{3}\sin x$

3. $y = 1.5\sin x$

4. $y = -5\sin x$

5. $y = \sqrt{2}\sin x$

For questions 6 to 10, answer as indicated.

6. What is the maximum value of $y = -10\sin x$?

7. What is the minimum value of $y = \dfrac{4}{5}\sin x$?

8. For what values of x in the interval $0 \le x \le 2\pi$ is $1.5\sin x = 0$?

9. For what values of x in the interval $0 \le x \le 2\pi$ is $y = -\sqrt{3}\sin x$ a maximum value?

10. For what values of x in the interval $0 \le x \le 2\pi$ is $y = -5\sin x$ a minimum value?

For questions 11 to 15, sketch one cycle of the graph of the given function.

11. $y = 4\sin x$

12. $y = 1.5\sin x$

13. $y = -5\sin x$

14. $y = \dfrac{4}{5}\sin x$

15. $y = -\sqrt{3}\sin x$

EXERCISE 9-3

For questions 1 to 5, state the period of the given function.

1. $y = \sin 4x$

2. $y = -5\sin 6x$

3. $y = \sqrt{2}\sin\dfrac{x}{3}$

4. $y = -0.6\sin\dfrac{1}{2}x$

5. $y = \dfrac{4}{5}\sin\pi x$

For questions 6 to 10, answer as indicated.

6. For what values of x in the interval $0 \le x \le 2\pi$ is $\sin 4x = 0$?

7. For what values of x in the interval $0 \le x \le 2\pi$ is $y = \sin 4x$ a maximum?

8. For what values of x in the interval $0 \le x \le 2\pi$ is $y = \sin 4x$ a minimum?

9. For what values of x in the interval $0 \le x \le 4\pi$ is $-0.6\sin\dfrac{1}{2}x = 0$?

10. For what value of x in the interval $0 \le x \le 4\pi$ is $y = -0.6\sin\dfrac{1}{2}x$ a maximum?

For questions 11 to 15, sketch at least one cycle of the graph of the given function.

11. $y = -0.6\sin\dfrac{1}{2}x$

12. $y = \sin 4x$

13. $y = -1\sin 2.5x$

14. $y = \sqrt{2}\sin\dfrac{x}{3}$

15. $y = \dfrac{4}{5}\sin\pi x$

EXERCISE 9-4

For questions 1 to 5, use the equations $x - \dfrac{C}{B} = 0$ and $x - \dfrac{C}{B} = \dfrac{2\pi}{|B|}$ to determine a one-cycle interval for the graph of the given function.

1. $y = -3\sin\left(-\dfrac{2x}{3} + \dfrac{\pi}{4}\right)$

2. $y = -2\sin\left(-2x + \dfrac{\pi}{2}\right)$

3. $y = \dfrac{1}{2}\sin(-2x + 4)$

4. $y = 3\sin(-2\pi x + \pi)$

5. $y = \dfrac{3}{4}\sin\left(-x - \dfrac{\pi}{3}\right)$

For questions 6 to 10, describe the major features of the graph of the given function and sketch at least one cycle of the function's graph.

6. $y = -3\sin\left(-\dfrac{2x}{3} + \dfrac{\pi}{4}\right)$

7. $y = -2\sin\left(-2x + \dfrac{\pi}{2}\right)$

8. $y = \dfrac{1}{2}\sin(-2x + 4)$

9. $y = 3\sin(-2\pi x + \pi)$

10. $y = \dfrac{3}{4}\sin\left(-x - \dfrac{\pi}{3}\right)$

EXERCISE 9-5

For questions 1 to 4, describe the major features of the graph of the given function and sketch at least one cycle of the function's graph.

1. $y = 5\sin\left(x - \dfrac{\pi}{4}\right) - 1$

2. $y = 2\sin\left(x + \dfrac{\pi}{4}\right) + 3$

3. $y = -\sin\left(-2x + \dfrac{\pi}{3}\right) + \dfrac{1}{2}$

4. $y = \dfrac{1}{2}\sin(-3x - \pi) + 3$

One more question!

5. The current I (in amperes) in a wire of an alternating-current circuit is given by the function $I(t) = 15\sin 120\pi t$, where t is the elapsed time in seconds. What is the period of $I(t)$? What is the maximum value of the current?

Flashcard
App

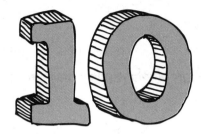

Graphs of the Cosine Function

MUST ⚡ KNOW

⚡ The graph of $y = \cos x$ provides a visual representation of its properties, including its period, amplitude, maximum value, minimum value, and midline.

⚡ The domain of $y = \cos x$ is $(-\infty, \infty)$, and its range is $[-1, 1]$. Its graph oscillates about the x-axis with a period of 2π, amplitude of 1, and x-intercepts at odd-integer multiples of $\frac{\pi}{2}$.

⚡ Under transformations, the graph of the cosine function can be stretched or compressed vertically or horizontally, shifted left or right or up or down, and reflected about a horizontal midline.

⚡ The graph of a cosine function provides a model for oscillating, harmonic, or wave motion.

I t's important for us to be knowledgeable about the graph of the cosine function and have a strong grasp of the impact on the graph when elements of its equation are altered. In this chapter, we're going to study the properties of the graph of the basic cosine function $y = \cos x$ as well as analyze the graph of $y = A\cos(Bx - C) + D$.

The Graph of $y = \cos x$

To graph the function $y = \cos x$, create an x-y table using some familiar values of x in the interval from 0 to 2π. Plot the table's ordered pairs, and then connect the points with a smooth curve.

x	0	$\dfrac{\pi}{4}$	$\dfrac{\pi}{3}$	$\dfrac{\pi}{2}$	$\dfrac{2\pi}{3}$	$\dfrac{3\pi}{4}$
$\cos x$	1	$\dfrac{\sqrt{2}}{2}$	$\dfrac{1}{2}$	0	$-\dfrac{1}{2}$	$-\dfrac{\sqrt{2}}{2}$
$(x,\cos x)$	$(0,1)$	$\left(\dfrac{\pi}{4},\dfrac{\sqrt{2}}{2}\right)$	$\left(\dfrac{\pi}{3},\dfrac{1}{2}\right)$	$\left(\dfrac{\pi}{2},0\right)$	$\left(\dfrac{2\pi}{3},-\dfrac{1}{2}\right)$	$\left(\dfrac{3\pi}{4},-\dfrac{\sqrt{2}}{2}\right)$
x	π	$\dfrac{5\pi}{4}$	$\dfrac{4\pi}{3}$	$\dfrac{3\pi}{2}$	$\dfrac{5\pi}{3}$	$\dfrac{7\pi}{4}$	2π
$\cos x$	-1	$-\dfrac{\sqrt{2}}{2}$	$-\dfrac{1}{2}$	0	$\dfrac{1}{2}$	$\dfrac{\sqrt{2}}{2}$	1
$(x,\cos x)$	$(\pi,-1)$	$\left(\dfrac{5\pi}{4},-\dfrac{\sqrt{2}}{2}\right)$	$\left(\dfrac{4\pi}{3},-\dfrac{1}{2}\right)$	$\left(\dfrac{3\pi}{2},0\right)$	$\left(\dfrac{5\pi}{3},\dfrac{1}{2}\right)$	$\left(\dfrac{7\pi}{4},\dfrac{\sqrt{2}}{2}\right)$	$(2\pi,1)$

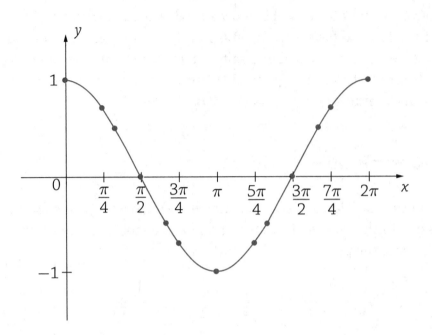

The graph shown represents only a portion of $y = \cos x$. This portion is the **basic cycle** of $y = \cos x$. The entire graph repeats the pattern of the basic cycle joined end to end in both the positive and negative directions on the x-axis. Following is a graph of $y = \cos x$ that shows several cycles, with the basic cycle plotted with a thicker line style.

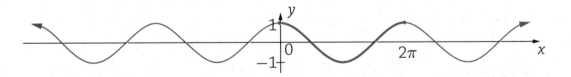

Like the sine function, the curve is smooth, continuous, and sinusoidal with no vertical asymptotes, holes, gaps, jumps, or corners.

The properties of $y = \cos x$ mimic those of $y = \sin x$. The domain of $y = \cos x$ is all real numbers. We can easily see in the following figure that its graph has an amplitude of 1 and a period of 2π. Its maximum value is 1, its minimum value is -1, and its range is $-1 \leq y \leq 1$.

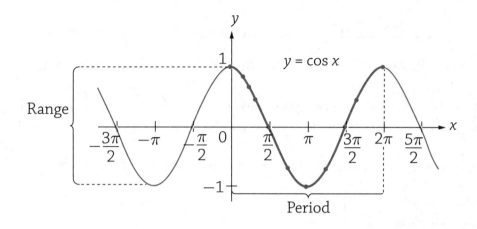

The real zeros of the cosine function, meaning the real numbers for which $\cos x = 0$, are odd-integer multiples of $\frac{\pi}{2}$: $\pm\frac{\pi}{2}$, $\pm\frac{3\pi}{2}$, $\pm\frac{5\pi}{2}$, and so on. This information tells us that the graph of $y = \cos x$ has x-intercepts at $\pm\frac{n\pi}{2}$, where n is an odd integer.

> **BTW**
>
> *The cosine function is an even function, meaning $\cos(-x) = \cos x$. Graphically, an even function is symmetric about the y-axis, as exhibited in the graph of $y = \cos x$.*

EXAMPLE

▶ For what values of x in the interval $0 \le x \le \pi$ is $y = \cos x$ decreasing?

▶ The graph of $y = \cos x$ is decreasing for x in the interval $0 < x < \pi$.

The Graph of $y = A\cos(Bx - C) + D$

The function $y = A\cos(Bx - C) + D = A\cos B\left(x - \frac{C}{B}\right) + D$ is a transformation of the function $y = \cos x$. Not surprisingly, the coefficients A, B, C, and D have the same effects on the cosine function as they did in the previous chapter on the sine function. Thus, the graph of $y = A\cos(Bx - C) + D$ has an amplitude of $|A|$ and a period of $\frac{2\pi}{|B|}$. The number $\frac{C}{B}$ induces a horizontal phase shift, with a shift to the right when $\frac{C}{B} > 0$, and to the left when $\dfrac{C}{B} < 0$. The number D induces a vertical shift that is D units upward when $D > 0$, and $|D|$ units downward when $D < 0$. The midline of the graph is $y = D$.

If $A < 0$, the graph is a reflection over the line $y = D$ of the corresponding function with $A > 0$. If $B < 0$, we can use the even function property of the cosine function to obtain the equivalent function $y = A\cos\left(-|B|\left(x + \dfrac{C}{|B|}\right)\right) +$ $D = A\cos\left(|B|\left(x + \dfrac{C}{|B|}\right)\right) + D.$

> 🌐 **IRL** In engineering, two periodic functions with the same period are *out of phase* if their phase shifts are different. The sine and cosine functions are out of phase by $\dfrac{\pi}{2}$ units. Their graphs coincide when the graph of the sine function is shifted left by $\dfrac{\pi}{2}$ units, or when the graph of the cosine function is shifted right by the same amount. That is, $\sin\left(x + \dfrac{\pi}{2}\right) = \cos x$ and $\cos\left(x - \dfrac{\pi}{2}\right) = \sin x.$

Here is an example in which the midline is above the x-axis.

EXAMPLE

▶ Describe the major features of the graph of $y = 3\cos\left(x + \dfrac{\pi}{6}\right) + 2$ and sketch at least one cycle of the graph.

▶ The graph of $y = 3\cos\left(x + \dfrac{\pi}{6}\right) + 2 = 3\cos\left(x - \left(-\dfrac{\pi}{6}\right)\right) + 2$ has period 2π, amplitude 3, a phase shift of $\dfrac{\pi}{6}$ to the left, a vertical shift of 2 upward, and a midline equation of $y = 2$.

▶ Solving $x - \left(-\dfrac{\pi}{6}\right) = 0$ and $x - \left(-\dfrac{\pi}{6}\right) = 2\pi$ yields $\left[-\dfrac{\pi}{6}, \dfrac{11\pi}{6}\right]$ as a one-cycle interval of the graph. The following graph of $y = 3\cos\left(x + \dfrac{\pi}{6}\right) + 2 = 3\cos\left(x - \left(-\dfrac{\pi}{6}\right)\right) + 2$, in which the midline equation $y = 2$ is plotted in grayscale as a dashed line and the portion of the graph corresponding to the interval $\left[-\dfrac{\pi}{6}, \dfrac{11\pi}{6}\right]$ is plotted in a thicker line style.

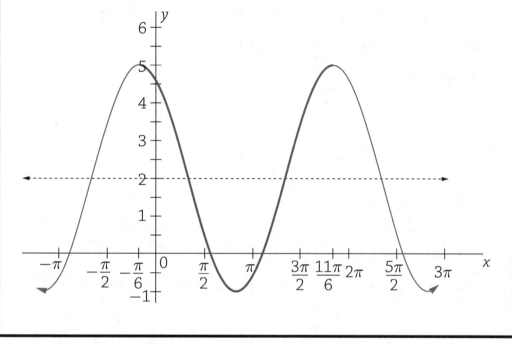

Let's work with an example in which the midline is below the x-axis.

EXAMPLE

▶ Describe the major features of the graph of $y = 2\cos\left(4x - \dfrac{4\pi}{3}\right) - 3$ and sketch at least one cycle of the graph.

▶ The function $y = 2\cos\left(4x - \dfrac{4\pi}{3}\right) - 3 = 2\cos\left(4\left(x - \dfrac{\pi}{3}\right)\right) - 3$ has period $\dfrac{\pi}{2}\left(= \dfrac{2\pi}{4}\right)$, amplitude 2, a phase shift of $\dfrac{\pi}{3}$ to the right, a vertical shift of 3 downward, and a midline equation of $y = -3$.

▶ Solving $x - \dfrac{\pi}{3} = 0$ and $x - \dfrac{\pi}{3} = \dfrac{\pi}{2}$ yields $\left[\dfrac{\pi}{3}, \dfrac{5\pi}{6}\right]$ as a one-cycle interval of the graph. The following graph of

$y = 2\cos\left(4x - \dfrac{4\pi}{3}\right) - 3 = 2\cos\left(4\left(x - \dfrac{\pi}{3}\right)\right) - 3$, in which the midline

equation $y = -3$ is plotted in grayscale as a dashed line and the

portion of the graph corresponding to the interval $\left[\dfrac{\pi}{3}, \dfrac{5\pi}{6}\right]$ is plotted in

a thicker line style.

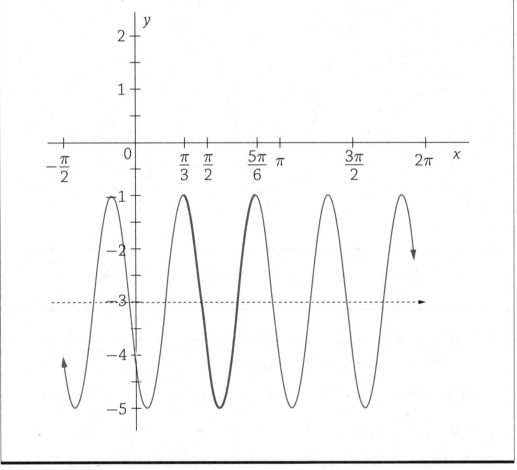

The graph in the following example confirms that the graph of $y = \cos x$

shifted $\dfrac{\pi}{2}$ to the right is identical to the graph of $y = \sin x$.

EXAMPLE

▶ Describe the major features of the graph of $y = \cos\left(x - \dfrac{\pi}{2}\right)$ and sketch at least one cycle of the graph.

▶ The function $y = \cos\left(x - \dfrac{\pi}{2}\right)$ has period 2π, amplitude 1, a phase shift of $\dfrac{\pi}{2}$ to the right, no vertical shift, and a midline equation of $y = 0$.

▶ Solving $x - \dfrac{\pi}{2} = 0$ and $x - \dfrac{\pi}{2} = 2\pi$ yields $\left[\dfrac{\pi}{2}, \dfrac{5\pi}{2}\right]$ as a one-cycle interval of the graph. The following graph of $y = \cos\left(x - \dfrac{\pi}{2}\right)$, in which the portion of the graph corresponding to the interval $\left[\dfrac{\pi}{2}, \dfrac{5\pi}{2}\right]$ is plotted in a thicker line style.

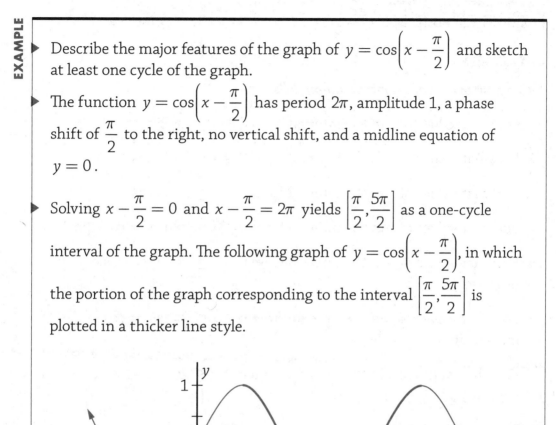

Observe that the previous graph of $y = \cos\left(x - \dfrac{\pi}{2}\right)$ is identical to the graph of $y = \sin x$ in Chapter 9.

EXERCISES

EXERCISE 10-1

Let's try out our new cos function-graph skills.

1. For what values of x in the interval from $0 \le x \le 2\pi$ is $\cos x = 0$?

2. For what values of x in the interval from $0 \le x \le 2\pi$ is $\cos x = 1$?

3. For what values of x in the interval from $0 \le x \le 2\pi$ is $\cos x = -1$?

4. For what values of x in the interval from $0 \le x \le 2\pi$ is $y = \cos x$ positive?

5. For what values of x in the interval from $0 \le x \le 2\pi$ is $y = \cos x$ negative?

6. For what values of x in the interval from $0 \le x \le 2\pi$ is $y = \cos x$ increasing?

7. For what values of x in the interval from $0 \le x \le 2\pi$ is $y = \cos x$ decreasing?

8. Is the value -5 in the domain of $y = \cos x$? Yes or No? Justify your answer.

9. Is the value -5 in the range of $y = \cos x$? Yes or No? Justify your answer.

10. Is $\cos(-x) = -\cos x$? Yes or No? Justify your answer.

EXERCISE 10-2

Describe the major features of each graph of the given function and sketch at least one cycle of the graph.

1. $y = 5\cos\left(x - \dfrac{\pi}{4}\right) - 1$

2. $y = 2\cos\left(x + \dfrac{\pi}{4}\right) + 3$

3. $y = -\cos\left(-2x + \dfrac{\pi}{3}\right) + \dfrac{1}{2}$

4. $y = \dfrac{1}{2}\cos(-3x - \pi) + 3$

One more!

5. The y-coordinate of a circular path centered at the origin that models an object traveling at constant angular speed is given by the function $y = 10\cos\left(\pi t - \dfrac{\pi}{12}\right)$, where t is the time in seconds. What is the period of the function? What is the maximum value of y?

Flashcard App

Graphs of the Tangent Function

MUST KNOW

- The graph of $y = \tan x$ provides a visual representation of its properties, including its period, midline, x-intercepts, and vertical asymptotes.

- The domain of $y = \tan x$ is all real numbers except for odd multiples of $\frac{\pi}{2}$, and its range is $(-\infty, \infty)$. Its graph has period of π, x-intercepts at integer multiples of π, and vertical asymptotes at odd-integer multiples of $\frac{\pi}{2}$.

- Under transformations, the graph of the tangent function can be stretched or compressed vertically or horizontally, shifted left or right or up or down, and reflected about a horizontal midline.

We're now going to get acquainted with significant features of the graphs of the tangent function and of its variations. We'll explore the properties of the graph of the basic tangent function $y = \tan x$ and then analyze the graph of $y = A\tan(Bx - C) + D$.

The Graph of $y = \tan x$

By definition, $\tan x = \dfrac{\sin x}{\cos x}$. (See Chapter 7 for a review.) Therefore, the domain of the tangent function is the set of real numbers except those numbers for which $\cos x = 0$. Thus, for the function $y = \tan x$, we specify that $x \neq \dfrac{n\pi}{2}$, where n is an odd integer. The graph of $y = \tan x$ has vertical asymptotes at these excluded values of x. The x-intercepts of $y = \tan x$ occur at $x = n\pi$, where n is any integer. The range of $y = \tan x$ is R, the set of real numbers. The function has period π. Its graph has undefined amplitude because it has neither a maximum value nor a minimum value.

Following is a graph of $y = \tan x$ that shows several cycles, with the basic cycle in the interval $\left(-\dfrac{\pi}{2}, \dfrac{\pi}{2}\right)$ plotted with a thicker line style.

> **BTW**
>
> *The tangent function is odd, meaning $\tan(-x) = -\tan x$. Graphically, an odd function is symmetric about the origin, as exhibited in the graph of $y = \tan x$.*

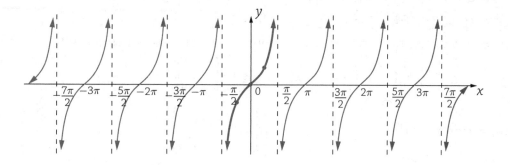

EXAMPLE

▶ The domain of the tangent function excludes which values?

▶ The domain of the tangent functions excludes values for which $\cos x = 0$. Thus, the tangent function's domain excludes $x = \dfrac{n\pi}{2}$, where n is an odd integer.

IRL The graph of the tangent function is used in electronics to illustrate and explain the workings of batteries and circuits and in seismic activity to investigate the formation of earthquakes and tsunamis.

The Graph of $y = A\tan(Bx - C) + D$

The function $y = A\tan(Bx - C) + D$ has domain $x \neq \dfrac{C}{B} + \dfrac{n\pi}{2B}$, where n is an odd integer, and its range is $(-\infty, \infty)$. Its graph has undefined amplitude, stretching (or compression) factor of $|A|$, period $\dfrac{\pi}{|B|}$, horizontal phase shift $\dfrac{C}{B}$, vertical phase shift D, and vertical asymptotes at $x = \dfrac{C}{B} + \dfrac{n\pi}{2B}$, where n is an odd integer.

If $A < 0$, the graph is a reflection over the line $y = D$ of the corresponding function with $A > 0$. If $B < 0$, we can use the odd function property of the tangent function to obtain the equivalent function

$$y = A\tan\left(-|B|\left(x + \dfrac{C}{|B|}\right)\right) + D = -A\tan\left(|B|\left(x + \dfrac{C}{|B|}\right)\right) + D, \text{ which is a}$$

reflection over the line $y = D$ of $A\tan\left(|B|\left(x + \dfrac{C}{|B|}\right)\right) + D$.

EXAMPLE

▶ Describe the major features of the graph of $y = 2\tan\left(\dfrac{x}{4}\right) - 3$ and sketch at least one cycle of the graph.

▶ The graph of $y = 2\tan\left(\dfrac{x}{4}\right) - 3$ shown next has period $4\pi \left(= \dfrac{\pi}{\frac{1}{4}} \right)$;

stretching factor 2; vertical asymptotes at $x = n \cdot 2\pi$, where n is an odd integer; and a vertical shift of 3 downward.

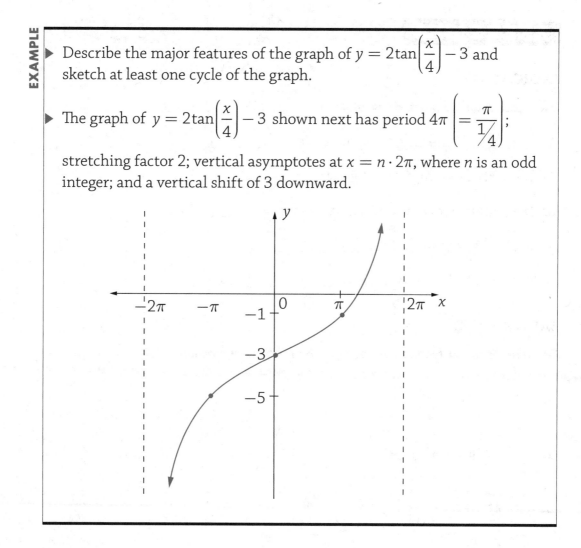

 **IRL** The cosecant, secant, and cotangent functions are the reciprocal functions of the sine, cosine, and tangent functions, respectively, and so are also periodic functions. Their graphs are not as useful and are seldom encountered in real-life applications—likely because, outside of pure mathematics, it's sufficient to use the graphs of sine, cosine, and tangent.

EXERCISES

EXERCISE 11-1

Fill in each blank to make a true statement.

1. The range of the tangent function is _____.

2. The vertical asymptotes of $y = \tan x$ occur at $x =$ _____.

3. The x-intercepts of $y = \tan x$ occur at $x =$ _____.

4. The period of $y = \tan x$ is _____.

5. The amplitude of $y = \tan x$ is _____.

EXERCISE 11-2

Describe the major features of each graph of the given function.

1. $y = 3\tan\left(x - \dfrac{\pi}{4}\right) - 2$

2. $y = -5\tan\left(x + \dfrac{\pi}{4}\right) + 1$

3. $y = 4\tan\left(2x + \dfrac{\pi}{6}\right) + \dfrac{1}{2}$

4. $y = \dfrac{1}{2}\tan(2x - \pi) + \sqrt{5}$

One more! *Describe the major features of the graph of the given function and sketch at least one cycle of the graph.*

5. $y = -\tan\left(-x + \dfrac{\pi}{3}\right)$

Flashcard App

Inverse Trigonometric Functions

MUST KNOW

⚡ None of the trigonometric functions are one-to-one functions on their domains; however, by restricting their domains, we can obtain one-to-one trigonometric functions that have inverse trigonometric functions.

⚡ The inverse sine function $y = \sin^{-1} x$ has domain $-1 \leq x \leq 1$ and range $-\frac{\pi}{2} \leq y \leq \frac{\pi}{2}$.

⚡ The inverse cosine function $y = \cos^{-1} x$ has domain $-1 \leq x \leq 1$ and range $0 \leq y \leq \pi$.

⚡ The inverse tangent function $y = \tan^{-1} x$ has domain $-\infty < x < \infty$ and range $-\frac{\pi}{2} < y < \frac{\pi}{2}$.

⚡ The inverse trigonometric functions allow us to find the angle measure or real number associated with the value of a trigonometric function.

U p to this point, we have learned to find the values of trigonometric functions of real numbers (or angles). In many applications, we must do the reverse. What do we do if we are given the value of a trigonometric function and we need to determine a corresponding real number (or angle) that has that trigonometric function value? This situation calls for the notion of inverse trigonometric functions.

Definitions and Basic Concepts of the Inverse Sine, Cosine, and Tangent Functions

To prepare ourselves for a discussion of the inverse trigonometric functions, we recall the following relationships between any function and its inverse:

Let f be a one-to-one function with domain X and range Y. The **inverse of f** is the unique function f^{-1} with domain Y and range X, such that $y = f(x)$ if and only if $x = f^{-1}(y)$. Equivalently, $f(f^{-1}(y)) = y$ for every $y \in Y$ if and only if $f^{-1}(f(x)) = x$ for every $x \in X$.

A function is **one-to-one** if every element of the range corresponds to exactly one element of the domain of the function. For a function to have an inverse function, the original function must be one-to-one. The trigonometric functions are periodic on the real numbers. Consequently, *none* of the trigonometric functions are one-to-one functions on their domains. However, opportunely, if the domain of a trigonometric function is suitably restricted to a portion on which the function *is* one-to-one, then it has a unique inverse function on that restricted domain. Moreover, a restricted domain should cover the entire range of the original function. So, let's proceed with that in mind.

Conventional choices for the restricted domains of the sine, cosine, and tangent functions are given in the following table.

Restricted Domains of the Sine, Cosine, and Tangent Functions

Function	Restricted Domain
$y = \sin x$	$-\dfrac{\pi}{2} \le x \le \dfrac{\pi}{2}$
$y = \cos x$	$0 \le x \le \pi$
$y = \tan x$	$-\dfrac{\pi}{2} < x < \dfrac{\pi}{2}$

Using the restricted domains from the previous table, the inverse sine, cosine, and tangent functions (represented using the notations $\sin^{-1} x$, $\cos^{-1} x$, and $\tan^{-1} x$, respectively) are defined so that their domains and ranges are as given in the following table.

Domains and Ranges of the Inverse Sine, Cosine, and Tangent Functions

Function	Domain	Range
$y = \sin^{-1} x$	$-1 \le x \le 1$	$-\dfrac{\pi}{2} \le y \le \dfrac{\pi}{2}$
$y = \cos^{-1} x$	$-1 \le x \le 1$	$0 \le y \le \pi$
$y = \tan^{-1} x$	$-\infty < x < \infty$	$-\dfrac{\pi}{2} < y < \dfrac{\pi}{2}$

Notice that the outputs of the inverse trigonometric functions are limited to specific range values. These limitations are a necessary result of the one-to-one function requirement for the function and its inverse.

BTW

In the notation $\sin^{-1} x$ (read as "inverse sine of x"), the small raised "−1" is not an exponent. Instead, it simply denotes an inverse function.

The graphs of the trigonometric sine, cosine, and tangent functions on their restricted domains and their corresponding inverses are shown next.

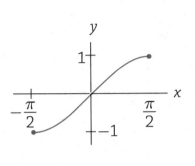

sin x

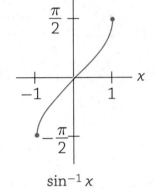

$\sin^{-1} x$

BTW

Other notations for the inverse trigonometric functions use an "arc-" prefix, e.g., arcsin x, read as "an angle whose sine is x."

BTW

The graph of a trigonometric function on its restricted domain and its inverse are reflections over the line y = x.

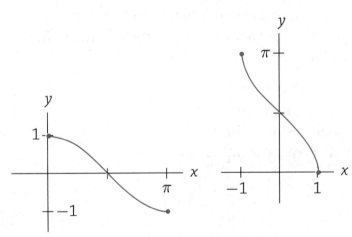

cos x

$\cos^{-1} x$

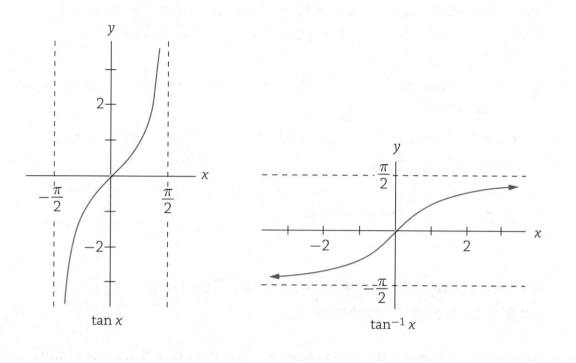

tan x

tan⁻¹ x

 IRL   Because the trigonometric functions are periodic, intervals other than those selected for their restricted domains could be chosen; however, mathematicians generally agree on the intervals given in this chapter, as evidenced by their adoption in technology.

Notice that the ranges of the trigonometric functions on their restricted domains are identical to their ranges on the real numbers. Therefore, you can enter any number in a trigonometric function's range as the input for the corresponding inverse trigonometric function. However, the output of the inverse trigonometric function must lie in a limited range interval that matches up with the restricted domain used to make the original function one-to-one. For example, $\sin^{-1}\left(\dfrac{1}{2}\right)$ is $\dfrac{\pi}{6}$ because $\dfrac{\pi}{6}$ is the one and only angle in the interval $\left[-\dfrac{\pi}{2}, \dfrac{\pi}{2}\right]$ whose sine is $\dfrac{1}{2}$. As explained in Appendix A, the trigonometric features on calculators use these limited ranges to determine output values for the inverse trigonometric functions. To summarize:

- $\sin^{-1} x$ is the real number (or angle) in the interval $\left[-\dfrac{\pi}{2}, \dfrac{\pi}{2}\right]$ whose sine is x, where $-1 \le x \le 1$.

- $\cos^{-1} x$ is the real number (or angle) in the interval $[0, \pi]$ whose cosine is x, where $-1 \le x \le 1$.

- $\tan^{-1} x$ is the real number (or angle) in the interval $\left(-\dfrac{\pi}{2}, \dfrac{\pi}{2}\right)$ whose tangent is x, where $-\infty < x < \infty$.

Evaluating the Inverse Sine, Cosine, and Tangent Functions

As we move into evaluating the inverse trigonometry functions, let's remind ourselves that, by definition, the inverse trigonometric functions return outputs that fall only in specified intervals. For example, even though we know that $\sin \dfrac{11\pi}{6} = -\dfrac{1}{2}$, it is not true that $\sin^{-1}\left(-\dfrac{1}{2}\right) = \dfrac{11\pi}{6}$. Why not? The simple answer is that the output of the inverse sine function must lie in the interval $\left[-\dfrac{\pi}{2}, \dfrac{\pi}{2}\right]$. Because $\dfrac{11\pi}{6}$ does not satisfy this requirement, $\sin^{-1}\left(-\dfrac{1}{2}\right) \ne \dfrac{11\pi}{6}$. So, what is the correct output for $\sin^{-1}\left(-\dfrac{1}{2}\right)$? Well, let's find a real number (or angle) that lies in the interval $\left[-\dfrac{\pi}{2}, \dfrac{\pi}{2}\right]$ and whose sine is $-\dfrac{1}{2}$. It turns out that there is only one possibility, which is $-\dfrac{\pi}{6}$. Therefore, $\sin^{-1}\left(-\dfrac{1}{2}\right) = -\dfrac{\pi}{6}$.

Accordingly, we cannot say too often that when you are evaluating $\sin^{-1} x$, $\cos^{-1} x$, and $\tan^{-1} x$, you need to be mindful of the following restrictions:

- For $\sin^{-1} x$, the output must lie in the interval $\left[-\dfrac{\pi}{2}, \dfrac{\pi}{2}\right]$.

- For $\cos^{-1} x$, the output must lie in the interval $[0, \pi]$.

- For $\tan^{-1} x$, the output must lie in the interval $\left(-\dfrac{\pi}{2}, \dfrac{\pi}{2}\right)$.

In this chapter, when you evaluate $\sin^{-1} x$, $\cos^{-1}$ x, or $\tan^{-1} x$, express the result in radians. If you are asked for an exact value of an inverse trigonometric function, put your calculator aside. You should anticipate that your answer is either a special angle or a quadrantal angle. In such cases, express the special angle or quadrantal angle as radians in terms of π. (For a helpful reference, see the table titled "Trigonometric Function Values of Special Angles and Quadrantal Angles" in Chapter 6.)

BTW

Note that none of the inverse trigonometric functions returns an output that lies in the interval $\left(\pi, \dfrac{3\pi}{2}\right)$, which is quadrant III.

Let's try an example that involves the inverse sine function.

EXAMPLE

▶ Find the exact value of $\sin^{-1} \dfrac{\sqrt{2}}{2}$. Do not use a calculator.

▶ Because $\sin \dfrac{\pi}{4} = \dfrac{\sqrt{2}}{2}$ and $\dfrac{\pi}{4}$ lies in the interval $\left[-\dfrac{\pi}{2}, \dfrac{\pi}{2}\right]$, $\sin^{-1} \dfrac{\sqrt{2}}{2} = \dfrac{\pi}{4}$.

Next, let's do an example that involves the inverse cosine function.

▶ Find the exact value of $\cos^{-1}\left(-\dfrac{1}{2}\right)$. Do not use a calculator.

▶ Because $\cos\dfrac{2\pi}{3} = -\dfrac{1}{2}$ and $\dfrac{2\pi}{3}$ lies in the interval $[0,\ \pi]$,

$\cos^{-1}\left(-\dfrac{1}{2}\right) = \dfrac{2\pi}{3}$.

Now let's look at an example involving the inverse tangent function.

▶ Find the exact value of $\tan^{-1}(-1)$. Do not use a calculator.

▶ Because $\tan\left(-\dfrac{\pi}{4}\right) = -1$ and $-\dfrac{\pi}{4}$ lies in the interval $\left(-\dfrac{\pi}{2},\ \dfrac{\pi}{2}\right)$,

$\tan^{-1}(-1) = -\dfrac{\pi}{4}$.

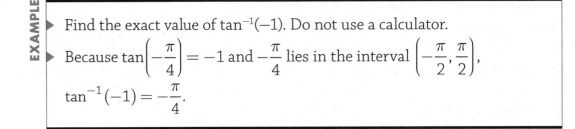

IRL Inverse trigonometric functions are widely used in engineering, navigation, physics, and astronomy to determine the measures of unknown angles.

One legendary example of an application of inverse trigonometric functions was in the determination of the angle of tilt of the Leaning Tower of Pisa, in the Tuscany region of Italy. The tower began to tilt during its construction in the twelfth century due to an inadequate foundation. Over time, the tilt increased and with a rate that indicated the tower eventually would topple over. In 1993, the angle of tilt of the tower was approximately 5.5°, which can be found by calculating the inverse sine of a ratio of measurements taken from the tower. In the next decade or so, an international team that included structural engineers, geo-technicians, and even historians worked to save the architectural treasure. Through their efforts, the lean was reduced to less than 4°, and the tower now is believed to be structurally stable for at least another 200 years.

Now let's try an example that involves evaluating the composition of trigonometric functions.

EXAMPLE

▶ Evaluate $\sin\left(\cos^{-1}\dfrac{\sqrt{3}}{2}\right)$. Do not use a calculator.

▶ $\sin\left(\cos^{-1}\dfrac{\sqrt{3}}{2}\right) = \sin\dfrac{\pi}{6} = \dfrac{1}{2}$

Stay sharp when doing the next example.

EXAMPLE

▶ Evaluate $\cos^{-1}\left(\cos\dfrac{5\pi}{3}\right)$. Do not use a calculator.

▶ $\cos^{-1}\left(\cos\dfrac{5\pi}{3}\right) = \cos^{-1}\left(\dfrac{1}{2}\right) = \dfrac{\pi}{3}$. This result is true because $\cos\dfrac{\pi}{3} = \dfrac{1}{2}$ and $\dfrac{\pi}{3}$ lies in the interval $[0,\pi]$.

▶ Be careful! $\cos^{-1}\left(\cos\dfrac{5\pi}{3}\right) \neq \dfrac{5\pi}{3}$ because $\dfrac{5\pi}{3}$ does *not* lie in the interval $[0, \pi]$.

BTW

Over the years, we've seen a lot of our students become frustrated after making errors such as

$\cos^{-1}\left(\cos\dfrac{5\pi}{3}\right) = \dfrac{5\pi}{3}$

or $\cos^{-1}\left(\dfrac{1}{2}\right) = \dfrac{5\pi}{3}$.

To guard against such mistakes, we strongly recommend committing to memory the restricted domains that are used to create the inverse trigonometric functions.

In the following example, we call on knowledge of the definitions of the trigonometric functions of any angle. (See Chapter 6 for a review.)

EXAMPLE

▶ Without using a calculator, find $\sin\left(\tan^{-1}\dfrac{3}{4}\right)$.

▶ Let $\theta = \tan^{-1}\dfrac{3}{4}$, then $\tan\theta = \dfrac{3}{4}$. We can situate θ as an angle in standard position in the coordinate plane, with (x, y) as a point on its terminal side.

▶ Using the definition that $\tan\theta = \dfrac{y}{x}$, we let $x = 4$ and $y = 3$,

yielding $r = \sqrt{4^2 + 3^2} = \sqrt{25} = 5$. By definition, $\sin\theta = \dfrac{y}{r} = \dfrac{3}{5}$.

Thus, $\sin\left(\tan^{-1}\dfrac{3}{4}\right) = \dfrac{3}{5}$.

This next example is more challenging.

EXAMPLE

▶ Write $\cos\left(\tan^{-1}\left(\dfrac{\sqrt{x^2 - 9}}{3}\right)\right)$ as an algebraic expression in x, where $x > 0$.

▶ Let $\theta = \tan^{-1}\left(\dfrac{\sqrt{x^2 - 9}}{3}\right)$. Then, $\tan\theta = \dfrac{\sqrt{x^2 - 9}}{3}$.

▶ Sketch a right triangle that fits this tangent ratio. Designate $\sqrt{x^2 - 9}$ as the side opposite θ and 3 as the side adjacent to θ, making the hypotenuse equal to $\sqrt{x^2 - 9 + 9} = \sqrt{x^2} = x$.

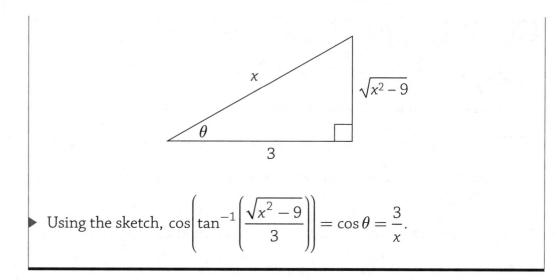

Using the sketch, $\cos\left(\tan^{-1}\left(\dfrac{\sqrt{x^2-9}}{3}\right)\right) = \cos\theta = \dfrac{3}{x}$.

In the next example, we're going to use a TI-84 Plus calculator to evaluate an inverse trigonometric function. (See Appendix A for instructions for determining inverse trigonometric values.) We do not have to check whether the calculator returns an output in the restricted range because the TI-84 Plus calculator, like most scientific and graphing calculators, is programmed to output answers consistent with the definitions of the inverse sine, cosine, and tangent functions given in this chapter.

EXAMPLE

Use a calculator to find an approximate real-number value for $\tan^{-1}(-0.5906)$. Round the answer to three decimal places.

Using the calculator in radian mode, $\tan^{-1}(-0.5906) \approx -0.533$.

IRL We also can define inverses for the secant, cosecant, and cotangent functions on appropriate restricted domains in a manner similar to that used to define the inverses of the sine, cosine, and tangent functions. However, the inverse secant, inverse cosecant, and inverse cotangent functions are used infrequently in the real world. A major limitation for these inverse functions is that they cannot be evaluated directly on most calculators.

EXERCISES

EXERCISE 12-1

Indicate whether each statement is true or false.

1. All of the trigonometric functions are one-to-one on their domains.

2. The outputs of the inverse trigonometric functions are limited to specific range values.

3. The inverse trigonometric functions are not periodic functions.

4. $\sin^{-1}\left(\dfrac{1}{2}\right) = \dfrac{\pi}{6}$.

5. Because $\tan\dfrac{5\pi}{4} = 1$, $\tan^{-1}(1) = \dfrac{5\pi}{4}$.

EXERCISE 12-2

For questions 1 to 5, find the exact value of each expression. Do not use a calculator.

1. $\cos^{-1} 0$

2. $\sin^{-1}\left(-\dfrac{\sqrt{2}}{2}\right)$

3. $\tan^{-1}\dfrac{1}{\sqrt{3}}$

4. $\sin^{-1}\left(-\dfrac{1}{2}\right)$

5. $\cos^{-1}\left(-\dfrac{\sqrt{3}}{2}\right)$

For questions 6 to 10, find the exact value of the expression. Do not use a calculator.

6. $\sin\left(\sin^{-1}\dfrac{\sqrt{3}}{2}\right)$

7. $\tan\left(\cos^{-1}\dfrac{1}{2}\right)$

8. $\sin^{-1}\left(\sin\dfrac{3\pi}{4}\right)$

9. $\cos^{-1}\left(\sin\dfrac{\pi}{6}\right)$

10. $\cos\left(\tan^{-1}\sqrt{3}\right)$

For questions 11 to 13, find the exact value of the expression. Do not use a calculator.

11. $\sin\left(\cos^{-1}\dfrac{24}{25}\right)$

12. $\cos\left(\tan^{-1}\dfrac{5}{12}\right)$

13. $\tan\left(\sin^{-1}\dfrac{4}{5}\right)$

For questions 14 and 15, write the given expression as an algebraic expression in x, where x > 0.

14. $\sin\left(\tan^{-1}\left(\dfrac{4}{\sqrt{x^2-16}}\right)\right)$

15. $\sin\left(\cos^{-1}\left(\dfrac{\sqrt{x^2-9}}{x}\right)\right)$

For questions 16 to 20, use a calculator in radian mode to find an approximate real-number value for the given expression. Round the answer to three decimal places.

16. $\sin^{-1}(0.3524)$

17. $\cos^{-1}(0.6705)$

18. $\tan^{-1}(-1.834)$

19. $\sin^{-1}(-0.9834)$

20. $\cos^{-1}(0.8962)$

Flashcard App

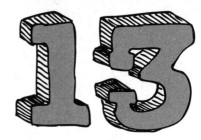

Solving Trigonometric Equations

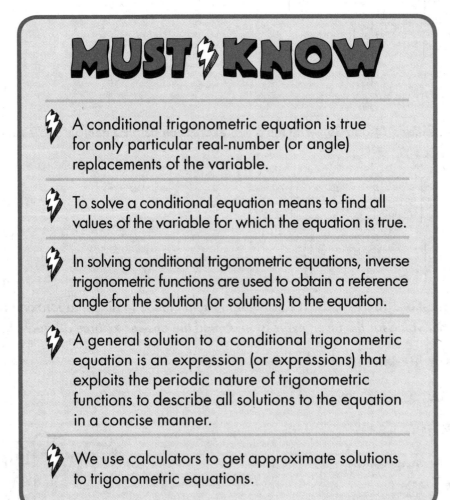

MUST KNOW

⚡ A conditional trigonometric equation is true for only particular real-number (or angle) replacements of the variable.

⚡ To solve a conditional equation means to find all values of the variable for which the equation is true.

⚡ In solving conditional trigonometric equations, inverse trigonometric functions are used to obtain a reference angle for the solution (or solutions) to the equation.

⚡ A general solution to a conditional trigonometric equation is an expression (or expressions) that exploits the periodic nature of trigonometric functions to describe all solutions to the equation in a concise manner.

⚡ We use calculators to get approximate solutions to trigonometric equations.

A trigonometric equation is an equation in which the variable is the argument of one or more trigonometric functions. In Chapter 7, we learned about trigonometric identities, which are equations that are true for all real-number (or angle) replacements of the variable for which all functions involved are defined. In this chapter, we learn about **conditional trigonometric equations**. These equations are true for only particular real-number (or angle) replacements of the variable. To unravel such an equation, we need techniques for finding those values, if any, of the variables that make the equation true.

We do not have a general procedure for solving all trigonometric equations. In this chapter, we consider only equations that can be solved by applying algebraic methods, such as rearranging or combining terms, reducing fractions, factoring, squaring, and taking roots in combination with using trigonometric identities to change the equation to a simpler form (e.g., $\sin x = k$, $\cos x = k$, or $\tan x = k$).

Here are general guidelines for solving such conditional trigonometric equations.

1. If the equation contains different trigonometric functions, use identities and algebraic manipulation (reducing fractions, adding/subtracting fractions, factoring, etc.) to express the different functions in terms of the same function.

2. If the arguments of the trigonometric functions involve an unknown angle and its multiples, use identities (e.g., double-angle, half-angle, and sum/difference formulas) to write all functions in terms of the same argument.

3. After all trigonometric functions are expressed in terms of the same trigonometric function with the same argument, isolate the remaining single trigonometric function to one side of the equation using algebraic techniques. Proviso: If the equation has a quadratic form, first use factoring or the quadratic formula to solve it for the trigonometric function involved.

4. Use your understanding of inverse trigonometric functions to obtain a reference angle for the variable.

5. Use the reference angle to determine all solutions in a specified interval or to write a general solution using your knowledge of the periodicity of trigonometric functions. A **general solution** to a trigonometric equation is an expression (or expressions) that describes all solutions to a trigonometric equation in a concise manner.

BTW

*When no confusion will result, we use the term **reference angle**, whether the argument of the function is a real number or an angle measured in degrees. A reference angle is always positive and less than $\frac{\pi}{2}$ or $90°$. See Chapter 1 for a full discussion of reference angles.*

Keep in mind that not all trigonometric equations have solutions, but solving those that do requires applying your knowledge of trigonometry from previous chapters in this book in combination with previously acquired algebraic manipulation skills.

The following table showing the algebraic signs of the trigonometric functions in each quadrant is a valuable tool to help in solving trigonometric equations.

Signs of the Trigonometric Functions in Quadrants I to IV, $x \neq 0$, and $y \neq 0$

Function	Quadrant			
	I	**II**	**III**	**IV**
$y = \sin x$	+	+	−	−
$y = \cos x$	+	−	−	+
$y = \tan x$	+	−	+	−

If $x = 0$ or $y = 0$, then the angles involved are quadrantal angles and the trigonometric functions take on the special values of 0, 1, −1, or are undefined. We will draw on knowledge of the function values of quadrantal angles to solve equations involving quadrantal angles as special cases. (See Chapter 6 to review.)

Solving for Exact Solutions to Trigonometric Equations

We do not use calculators when solving for exact solutions to trigonometric equations. Instead, we rely on a good working knowledge of the trigonometric values of the special angles. (For a helpful reference, see the table titled "Trigonometric Function Values of Special Angles and Quadrantal Angles" in Chapter 6.)

We can solve trigonometric equations for exact real-number solutions, as shown in the following example.

▶ Find all real numbers satisfying $\sin x = \dfrac{1}{2}$. Do not use a calculator.

▶ Given $\sin x = \dfrac{1}{2}$, $\sin^{-1}\dfrac{1}{2} = \dfrac{\pi}{6}$ yields $\dfrac{\pi}{6}$ as the reference angle.
We know that the sine function is positive in quadrants I and II.
The only solutions between 0 and 2π are $x = \dfrac{\pi}{6}$ (quadrant I) and
$x = \pi - \dfrac{\pi}{6} = \dfrac{5\pi}{6}$ (quadrant II).

▶ Then, because the sine function is periodic with a period of 2π, all

real-number solutions of $\sin x = \dfrac{1}{2}$ are given by $x = \dfrac{\pi}{6} + n \cdot 2\pi$ or

$x = \dfrac{5\pi}{6} + n \cdot 2\pi$, where n is an integer.

In the next example, we solve a trigonometric equation for exact solutions expressed as angles in degrees.

▶ Determine all solutions to $\cos\theta = \dfrac{\sqrt{2}}{2}$, where θ is an angle measured in degrees. Do not use a calculator.

▶ Given $\cos\theta = \dfrac{\sqrt{2}}{2}$, $\cos^{-1}\dfrac{\sqrt{2}}{2} = 45°$, which yields
45° as the reference angle. We know that the cosine
function is positive in quadrants I and IV. The only
solutions between 0° and 360° are $\theta = 45°$
(quadrant I) and $\theta = 360° - 45° = 315°$ (quadrant IV).

▶ Then, because the cosine function is periodic with a

period of 360°, all solutions of $\cos\theta = \dfrac{\sqrt{2}}{2}$ are given

by $\theta = 45° + n \cdot 360°$ or $\theta = 315° + n \cdot 360°$, where
n is an integer.

BTW

Note that to determine all solutions of a trigonometric equation, we write a general solution in which we add multiples of the period to the solutions that occur in one cycle of the trigonometric function.

Hereafter, we use x to denote an argument of a trigonometric function that is a real number and θ to denote an argument that is an angle measured in degrees.

Let's try an example involving the tangent function.

EXAMPLE

▶ Determine all solutions to $\sqrt{3}\tan\theta + 1 = 0$. Do not use a calculator.

▶ First, isolate $\tan\theta$ to one side of the equation.

$$\sqrt{3}\tan\theta + 1 = 0$$
$$\sqrt{3}\tan\theta = -1$$
$$\tan\theta = -\frac{1}{\sqrt{3}}$$

▶ Given $\tan\theta = -\frac{1}{\sqrt{3}}$, $\tan^{-1}\left(-\frac{1}{\sqrt{3}}\right) = -30°$, which yields $30°$ as the reference angle. We know that the tangent function is negative in quadrants II and IV. The only solutions between $0°$ and $360°$ are $\theta = 180° - 30° = 150°$ (quadrant II) and $\theta = 360° - 30° = 330°$ (quadrant IV).

▶ Because the tangent function is periodic with a period of $180°$, all solutions of $\sqrt{3}\tan\theta + 1 = 0$ are given by $\theta = 150° + n \cdot 180°$ or $\theta = 330° + n \cdot 180°$, where n is an integer.

In the next example, we solve a trigonometric equation in a specified interval.

EXAMPLE

▶ Solve $\tan\theta = 1$ in the interval $0° \le \theta < 360°$. Do not use a calculator.

▶ Given $\tan\theta = 1$, $\tan^{-1}1 = 45°$, which yields a reference angle of $45°$. We know that the tangent function is positive in quadrants I and III.

> The only angles between $0°$ and $360°$ that satisfy the equation are $\theta = 45°$ (quadrant I) and $\theta = 180° + 45° = 225°$ (quadrant III). Thus, the two solutions to $\tan\theta = 1$ in the interval $0° \leq \theta \leq 360°$ are $\theta = 45°$ and $\theta = 225°$.

In the next example, we use an identity to transform the equation into a quadratic equation and then we solve the resulting equation by factoring.

EXAMPLE

> Solve $\tan x = \cot x$ in the interval $0 \leq x < 2\pi$. Do not use a calculator.

> First, substitute the reciprocal identity $\cot x = \dfrac{1}{\tan x}$ into the equation and then simplify.

$$\tan x = \cot x$$
$$\tan x = \frac{1}{\tan x}$$
$$\tan^2 x = 1$$
$$\tan^2 x - 1 = 0$$

> Now factor the resulting quadratic equation and solve for $\tan x$.

$$\tan^2 x - 1 = 0$$
$$\left(\tan x + 1\right)\left(\tan x - 1\right) = 0$$
$$\tan x = -1 \text{ or } \tan x = 1$$

> Finally, solve for x, if possible.

> If $\tan x = -1$, $\tan^{-1}(-1) = -\dfrac{\pi}{4}$ yields $\dfrac{\pi}{4}$ as a reference angle. We know the tangent function is negative in quadrants II and IV, so the only solutions for $\tan x = -1$ in $0 \leq x < 2\pi$ are $x = \dfrac{3\pi}{4}$ (quadrant II) and $x = \dfrac{7\pi}{4}$ (quadrant IV).

▶ If $\tan x = 1$, then $\tan^{-1} 1 = \dfrac{\pi}{4}$ yields $\dfrac{\pi}{4}$ as a reference angle. We know the tangent function is positive in quadrants I and III, so the only solutions for $\tan x = 1$ in $0 \le x < 2\pi$ are $x = \dfrac{\pi}{4}$ (quadrant II) and $x = \dfrac{5\pi}{4}$ (quadrant IV).

▶ Hence, the complete solution set in $0 \le x < 2\pi$ is given by

$$\left\{ \frac{\pi}{4}, \frac{3\pi}{4}, \frac{5\pi}{4}, \frac{7\pi}{4} \right\}.$$

Here is another example in which substituting an identity transforms the equation into a quadratic equation.

EXAMPLE

▶ Solve $\cos 2\theta - \sin \theta = 0$ in the interval $0 \le \theta < 90°$. Do not use a calculator.

▶ First, substitute the double-angle identity $\cos 2\theta = 1 - 2\sin^2\theta$ into the equation and then simplify.

$$\cos 2\theta - \sin \theta = 0$$
$$1 - 2\sin^2\theta - \sin \theta = 0$$
$$2\sin^2\theta + \sin \theta - 1 = 0$$

▶ Now factor the resulting quadratic equation and solve for $\sin \theta$.

$$2\sin^2\theta + \sin \theta - 1 = 0$$
$$(2\sin \theta - 1)(\sin \theta + 1) = 0$$
$$\sin \theta = \frac{1}{2} \text{ or } \sin \theta = -1$$

▶ Finally, solve for θ, if possible.

▶ If $\sin \theta = \dfrac{1}{2}$, $\sin^{-1}\left(\dfrac{1}{2}\right) = 30°$ implies $\theta = 30°$, which lies in the interval $0 \le \theta < 90°$.

▶ The equation $\sin\theta = -1$ is rejected because it does not have a solution for which θ lies in the interval $0 \le \theta < 90°$. Hence, the solution is $\theta = 30°$.

In the next example, we have an argument of $4x$. Let's see how to deal with this situation.

EXAMPLE

▶ Solve $\tan 4x = 1$ in the interval $0 \le x < \dfrac{\pi}{2}$. Do not use a calculator.

▶ If $\tan 4x = 1$, then $\tan^{-1} 1 = \dfrac{\pi}{4}$ yields $\dfrac{\pi}{4}$ as the reference angle. Then, because the tangent function is positive in quadrants I and III, $4x = \dfrac{\pi}{4}$ (quadrant I) or $4x = \dfrac{5\pi}{4}$ (quadrant III).

▶ Therefore, $x = \dfrac{\pi}{16}$ or $x = \dfrac{5\pi}{16}$, both of which lie in the interval $0 \le \theta \le \dfrac{\pi}{2}$. Hence, the complete solution set in $0 \le \theta \le \dfrac{\pi}{2}$ is given by $\left\{ \dfrac{\pi}{16}, \dfrac{5\pi}{16} \right\}$.

Be wary of making a mistake in the next example.

EXAMPLE

▶ Solve $\sin x \cos x = 2\cos x$ in the interval $0 \le x < 2\pi$. Do not use a calculator.

▶ Get all terms on one side of the equation and then factor. Do not divide both sides by $\cos x$—that's a mistake!

$$\sin x \cos x = 2\cos x$$
$$\sin x \cos x - 2\cos x = 0$$
$$\cos x (\sin x - 2) = 0$$

▶ Set each factor equal to zero, and solve for x, if possible.

▶ If $\cos x = 0$, then $x = \dfrac{\pi}{2}$ or $x = \dfrac{3\pi}{2}$. Because $\sin x = 2$ has no solution, the complete solution set in $0 \le x < 2\pi$ is given by $\left\{\dfrac{\pi}{2}, \dfrac{3\pi}{2}\right\}$.

> **BTW**
>
> *Be wary of dividing both sides of an equation by a variable expression, such as cos x, because you run the risk of dividing by zero.*

Solving for Approximate Solutions to Trigonometric Equations

With our calculators set to the desired mode, we use the inverse trigonometric function keys to get approximate solutions to trigonometric equations involving specific numerical values.

Recall that the inverse trigonometric functions have restricted outputs. Those restrictions are consistent with inverse function calculations with calculators. When the calculator returns a value, that output determines a reference angle that we can use to obtain the solution set of the equation. Generally, there will be two values between 0 and 2π (or 0° and 360°) that are possible solutions to the equation. (See Appendix A for a discussion of trigonometric calculator use.)

EXAMPLE

▶ Find all real numbers satisfying $\sin x = \dfrac{1}{4}$. Round answers to three decimal places.

▶ Given $\sin x = \dfrac{1}{4}$, $\sin^{-1}\dfrac{1}{4} \approx 0.253$ yields 0.253 as the approximate reference angle. We know that the sine function is positive in quadrants I and II. The only solutions between 0 and 2π are $x \approx 0.253$ (quadrant I) and $x \approx \pi - 0.253 = 2.889$ (quadrant II).

▶ Then, because the sine function is periodic with a period of 2π, all approximate real-number solutions of $\sin x = \dfrac{1}{4}$ are given by $x \approx 0.253 + n \cdot 2\pi$ or $x \approx 2.889 + n \cdot 2\pi$, where n is an integer.

In the next example, we find the solution in a specified interval.

EXAMPLE

▶ Find all real numbers satisfying $\cos 2x = \dfrac{2}{5}$ in the interval $0 \le x < \pi$. Round answers to three decimal places.

▶ Given $\cos 2x = \dfrac{2}{5}$, then $\cos^{-1} \dfrac{2}{5} \approx 1.159$ yields 1.159 as the approximate reference angle. We know that the cosine function is positive in quadrants I and IV.

▶ $2x \approx 1.159$ (quadrant I) or $2x \approx 2\pi - 1.159 = 5.124$ (quadrant IV).

▶ $x \approx 0.580$ or $x \approx 2.562$, both of which lie in the interval $0 \le x < \pi$.

▶ Hence, the two approximate solutions in $0 \le x < \pi$ are $x \approx 0.580$ and $x \approx 2.562$.

Let's try an example where the argument is an angle expressed in degrees.

EXAMPLE

▶ Solve $\sin \theta = 0.7346$ in the interval $0° \le \theta \le 360°$. Round answers to one decimal place.

▶ Given $\sin \theta = 0.7346$, $\sin^{-1}(0.7346) \approx 47.3°$ yields 47.3° as the reference angle.

▶ $\theta \approx 47.3°$ (quadrant I) or $\theta \approx 180° - 47.3° = 132.7°$ (quadrant II).

▶ Hence, the two approximate solutions in $0° \le \theta \le 360°$ are $\theta \approx 47.3°$ and $\theta \approx 132.7°$.

Let's look at an example in which the reference angle is not readily apparent.

EXAMPLE

▶ Solve $\cos\theta = -0.4356$ in $0° \leq \theta \leq 360°$. Round answers to one decimal place.

▶ Given $\cos\theta = -0.4356$, then $\cos^{-1}(-0.4356) \approx 115.8°$ yields a reference angle of $180° - 115.8° = 64.2°$. We know that the cosine function is negative in quadrants II and III.

▶ The approximate values for θ are $115.8°$ (quadrant II) and $180° + 64.2° = 244.2°$ (quadrant III).

▶ Hence, the two approximate solutions in $0° \leq \theta \leq 360°$ are $\theta \approx 115.8°$ and $\theta \approx 244.2°$.

IRL The hammer throw has been an event at the Summer Olympics since 1900. The "hammer" is actually a metal ball attached by a steel wire to a handle. The goal of the competition is to throw the hammer the farthest. Currently, the Olympic record for men is 84.80 meters, which was set by Sergey Litvinov of Russia in 1988, and the record for women is 82.29 meters, set by Anita Wlodarczyk of Poland in 2016.

The angle of release (measured from the horizon) of the hammer is a major factor that affects the distance of throw. We can solve the trigonometric equation $\cos 2\theta = \dfrac{\left(9.81\frac{m}{s^2}\right)h}{v^2 + \left(9.81\frac{m}{s^2}\right)h}$, where v is the velocity of release (in meters per second) and h is the height of release (in meters) to determine θ, the angle of release that will result in the maximum distance (in meters) for various combinations of v and h.

EXERCISES

EXERCISE 13-1

State whether the equation is a conditional equation (C) or an identity (I).

1. $2\cos x + 1 = 0$

2. $\tan\theta = \dfrac{\sin\theta}{\cos\theta}$

3. $\cos^2 x - \sin^2 x = 1$

4. $\tan 2x = \dfrac{2\tan x}{1 - \tan^2 x}$

5. $\sin x \tan \dfrac{x}{2} = \dfrac{2 - \sqrt{2}}{2}$

EXERCISE 13-2

Complete as indicated. Do not use a calculator.

1. Find all real numbers satisfying $2\cos x + 1 = 0$.

2. Determine all solutions to $\tan\theta = -\sqrt{3}$.

3. Solve $\cos^2 x - \sin^2 x = \dfrac{\sqrt{3}}{2}$ in $0 \le x \le \pi$.

4. Solve $\sin^2 x \cos x = \dfrac{1}{4}\cos x$ in $0 \le x \le 2\pi$.

5. Solve $\cos 3\theta = \dfrac{1}{2}$ in $0 \le \theta \le 120°$.

EXERCISE 13-3

Complete as indicated. Use a calculator set to the desired mode.

1. Find all real numbers satisfying $\sin x = 0.2419$. Round answers to three decimal places.

2. Solve $\cos \theta = 0.5150$ in $0° \leq \theta \leq 360°$. Round answers to one decimal place.

3. Solve $\cos \theta = -0.9085$ in $0° \leq \theta \leq 360°$. Round answers to one decimal place.

4. Solve $\sin 2x = -0.9781$ in $0 \leq x \leq \pi$. Round answers to three decimal places.

5. Use the formula $\cos 2\theta = \dfrac{\left(9.81 \frac{\text{m}}{\text{s}^2}\right)h}{v^2 + \left(9.81 \frac{\text{m}}{\text{s}^2}\right)h}$ to determine θ, the angle of release that will result in the maximum distance (in meters) for a hammer throw contestant who stands 1.8 meters tall and normally achieves a release velocity of $v = 13.8$ meters per second. Round the answer to one decimal place.

Flashcard App

Answer Key

Answers obtained using a calculator and rounded off may differ slightly from the rounded answers given in this answer key.

Just so you know, it is the custom of the authors to not round off after intermediate steps, but to wait, if possible, until the last calculation to round off answers.

Angles and Their Measure

EXERCISE 1-1

1. I
2. Negative *x*-axis
3. II
4. III
5. IV

EXERCISE 1-2

1. 50°
2. 74°
3. 2°
4. 36°

5. 43°

6. 60°

7. 158°

8. 99°

9. 90°

10. 79°

EXERCISE 1-3

1. 343°

2. 20°

3. 80°

4. 259°

5. 100°

EXERCISE 1-4

1. 30°

2. 45°

3. 60°

4. 17°

5. 75°

EXERCISE 1-5

1. $\dfrac{\pi}{6}$

2. $\dfrac{2\pi}{3}$

3. $\dfrac{\pi}{4}$

4. $\dfrac{\pi}{2}$

5. $\dfrac{3\pi}{4}$

6. $540°$

7. $225°$

8. $-45°$

9. $150°$

10. $405°$

11. $200.5°$

12. $80.2°$

13. $131.8°$

14. $263.6°$

15. $-143.2°$

EXERCISE 1-6

1. $s = r\theta = (18\,\text{ft})\left(\dfrac{\pi}{3}\right) = 6\pi\,\text{ft} \approx 18.8\,\text{ft}$

2. $s = r\theta = (12\,\text{m})\left(\dfrac{5\pi}{4}\right) = 15\pi\,\text{m} \approx 47.1\,\text{m}$

3. $A = \dfrac{1}{2}r^2\theta = \dfrac{1}{2}(18\,\text{ft})^2\left(\dfrac{\pi}{3}\right) \approx 169.6\,\text{ft}^2$

4. $A = \dfrac{1}{2}r^2\theta = \dfrac{1}{2}(12\,\text{m})^2\left(\dfrac{5\pi}{4}\right) \approx 282.7\,\text{m}^2$

5. $\dfrac{\left(960\,{}^{\text{rev}}\!/_{\text{min}}\right)\left(2\pi\,{}^{\text{radians}}\!/_{\text{rev}}\right)}{60\,{}^{\text{sec}}\!/_{\text{min}}} = 32\pi\ \text{radians per second}$

Concepts from Geometry

EXERCISE 2-1

1. 65°
2. 37°
3. 80°
4. 60°
5. 45°

EXERCISE 2-2

1. Yes, because $8 + 16 > 22$.
2. Yes, because $3 + 4 > 5$.
3. No, because $8 + 2 < 11$.
4. Yes, because $1 + 1 > 1$.
5. Yes, because $485 + 21 > 502$.

EXERCISE 2-3

1. 13
2. 24
3. 20
4. 65
5. Yes, because $13^2 + 84^2 = 85^2$.
6. Yes, because $39^2 + 80^2 = 89^2$.
7. No, because $17^2 + 41^2 \neq 55^2$.
8. 24 feet
9. ≈ 33.3 feet
10. 50 feet

EXERCISE 2-4

1. Yes, the ratios of corresponding sides are equal.

2. Yes, two angles of one triangle have the same measure as two angles of the other triangle.

3. No, because the sides have no common ratio.

4. 10

5. 8

Right Triangle Trigonometry

EXERCISE 3-1

1. $\dfrac{24}{25}$

2. $\dfrac{25}{7}$

3. $\dfrac{25}{24}$

4. $\dfrac{7}{25}$

5. $\dfrac{24}{7}$

6. $\dfrac{9}{41}$

7. $\dfrac{35}{12}$

8. $\dfrac{11}{61}$

9. $\dfrac{4}{3}$

10. $\dfrac{13}{12}$

EXERCISE 3-2

1. $1:\sqrt{3}:2$

2. $1:1:\sqrt{2}$

3. $60°$

4. $45°$

5. $45°$

EXERCISE 3-3

1. $\cos 30° = \dfrac{b}{26}$ yields $b = 13\sqrt{3}$

2. $\sin 60° = \dfrac{a}{48}$ yields $a = 24\sqrt{3}$

3. $\sin 30° = \dfrac{20}{c}$ yields $c = 40$

4. $\sin 45° = \dfrac{a}{10}$ yields $a = 5\sqrt{2}$

5. $\tan 60° = \dfrac{9}{b}$ yields $b = 3\sqrt{3}$

4

The Trigonometry of General Right Triangles

EXERCISE 4-1

1. $B = 77°, b \approx 43.3, c \approx 44.5$

2. $B = 62°, a \approx 11.2, c \approx 23.8$

3. $A = 73°, a \approx 13.1, c \approx 13.7$

4. $B = 60°, a \approx 46.2, c \approx 92.4$

5. $A \approx 36.9°, B \approx 53.1°, c = 5$

EXERCISE 4-2

1. $(40\,\text{ft})(\sin 60^\circ) \approx 34.6\,\text{ft}$

2. $\sin\theta = \dfrac{80}{140}; \theta \approx 34.8^\circ$

3. $(150\,\text{ft})(\tan 52^\circ) \approx 192.0\,\text{ft}$

4. $\tan 15^\circ = \dfrac{x+20}{100}; x \approx 6.8\,\text{ft}$

5. $(200\,\text{ft})(\sin 40^\circ) \approx 128.6\,\text{ft}$

The Trigonometry of Oblique Triangles

EXERCISE 5-1

1. $b = \sqrt{6^2 + 16^2 - 2(6)(16)\cos 60^\circ} = 14$

2. $\cos\theta = \dfrac{2^2 + (2\sqrt{2})^2 - (2\sqrt{5})^2}{2(2)(2\sqrt{2})}; \theta = 135^\circ$

3. $b = \sqrt{25^2 + 15^2 - 2(25)(15)\cos 110^\circ} \approx 33.3, A \approx 44.9^\circ, C \approx 25.1^\circ$

4. $\cos\theta = \dfrac{2.9^2 + 3.3^2 - 4.1^2}{2(2.9)(3.3)}; \theta \approx 82.5^\circ$

5. $\sqrt{2.6^2 + 4.3^2 - 2(2.6)(4.3)\cos 140^\circ} \approx 6.5$; thus, the resultant's magnitude ≈ 6.5 lb.

6. $\sqrt{8^2 + 20^2 - 2(8)(20)\cos 18^\circ} \approx 12.6$; thus, the distance ≈ 12.6 km.

7. $\sqrt{8^2 + 6^2 - 2(8)(6)\cos 113^\circ} \approx 11.7$; thus, the length ≈ 11.7 cm.

8. $a = 7, b = 5, c = 4; A \approx 101.5^\circ, B \approx 44.4^\circ, C \approx 34.1^\circ$

9. $\sqrt{500^2 + 600^2 - 2(500)(600)\cos 105°} \approx 874.8$; thus, the distance ≈ 874.8 km.

10. $\sqrt{1{,}500^2 + 2{,}000^2 - 2(1{,}500)(2{,}000)\cos 50°} \approx 1{,}547.0$; thus, the length $\approx 1{,}547.0$ ft.

EXERCISE 5-2

1. $a = \dfrac{\sqrt{3}\sin 45°}{\sin 30°} \approx 2.4$

2. $b = \dfrac{4\sin 60°}{\sin 75°} \approx 3.6$

3. $b = \dfrac{15\sin 45°}{\sin 30°} \approx 21.2$

4. $\sin\theta = \dfrac{10\sin 52°}{8}; \theta \approx 80.1°$ or $180° - 80.1° = 99.9°$

5. $\sin\theta = \dfrac{6\sin 30°}{2\sqrt{3}}; \theta = 60°$ or $180° - 60° = 120°$

6. $c = \dfrac{8\sin 40°}{\sin 33°} \approx 9.4$

7. $\dfrac{(50\,\text{lb})\sin 42°}{\sin 113°} \approx 36\,\text{lb}$

8. $d = \dfrac{(1{,}500\,\text{ft})(\sin 16°)}{\sin 42°}(\sin 58°) \approx 524.0\,\text{ft}$

9. $\dfrac{(2{,}500\,\text{ft})\sin 78.5°}{\sin 54.2°} \approx 3{,}020.5\,\text{ft}$

10. $h = \dfrac{(400\,\text{ft})(\sin 73°)}{\sin 56.6°}(\sin 50.4°) \approx 353.0\,\text{ft}$

EXERCISE 5-3

1. One

2. One

3. None

4. None

5. Two

6. One

7. One

8. None

9. None

10. One right triangle

EXERCISE 5-4

1. $B \approx 36.8°$

2. $b \approx 22.8$

3. $B \approx 77.7°$ or $B \approx 102.3°$

4. $b \approx 8.5$

5. $A \approx 51.9°$

6. $\sqrt{150^2 + 120^2 - 2(150)(120)\cos 85°} \approx 183.7$; thus, the distance ≈ 183.7 yd.

7. $(1,550 \text{ ft}) \tan 52° \approx 1,983.9 \text{ ft}$

8. $\dfrac{(400 \text{ ft})\sin 37°}{\sin 95°} \approx 241.6 \text{ ft}$

9. $\sqrt{18^2 + 23^2 - 2(18)(23)\cos 130°} \approx 37.2$; thus, the resultant's magnitude ≈ 37.2 lb.

10. $\dfrac{(24 \text{ m})(\sin 21°)}{\sin 14°}(\sin 35°) \approx 20.4 \text{ m}$

EXERCISE 5-5

1. $\dfrac{1}{2}(30)(16)\sin 26° \approx 105.2$

2. $\dfrac{1}{2}(25)(20)\sin 105° \approx 241.5$

3. $\dfrac{1}{2}(8 \text{ ft})(6 \text{ ft})\sin 54° \approx 19.4 \text{ ft}^2$

4. $\dfrac{1}{2}(10 \text{ in})(10 \text{ in})\sin 60° \approx 43.3 \text{ in}^2$

5. $6 \cdot \dfrac{1}{2}(3 \text{ cm})(3 \text{ cm})\sin 60° \approx 23.4 \text{ cm}^2$

6

Trigonometric Functions
of Any Angle

EXERCISE 6-1

1. $\sin\theta = -\dfrac{1}{5}$, $\cos\theta = \dfrac{2\sqrt{6}}{5}$, $\tan\theta = -\dfrac{1}{2\sqrt{6}}$

2. $\sin\theta = \dfrac{\sqrt{7}}{4}$, $\cos\theta = -\dfrac{3}{4}$, $\tan\theta = -\dfrac{\sqrt{7}}{3}$

3. $\sin\theta = -\dfrac{15}{17}$, $\cos\theta = -\dfrac{8}{17}$, $\tan\theta = \dfrac{15}{8}$

4. $\sec\theta = -\dfrac{13}{12}$, $\csc\theta = \dfrac{13}{5}$, $\cot\theta = -\dfrac{12}{5}$

5. $\sec\theta = \dfrac{25}{7}$, $\csc\theta = \dfrac{25}{24}$, $\cot\theta = \dfrac{7}{24}$

6. III

7. II

8. I

9. III

10. II

11. $\sin\theta = -\dfrac{12}{13}$, $\tan\theta = \dfrac{12}{5}$

12. $\cos\theta = \dfrac{8}{10}$, $\tan\theta = -\dfrac{6}{8}$

13. $\sin\theta = \dfrac{40}{41}$, $\cos\theta = \dfrac{9}{41}$

14. $\sin\theta = \dfrac{7}{25}$, $\cos\theta = -\dfrac{24}{25}$

15. $\cos\theta = -\dfrac{8}{17}$, $\tan\theta = \dfrac{15}{8}$

EXERCISE 6-2

1. $\theta = 90° - 54°$; $\theta = 36°$

2. $\theta = \dfrac{\pi}{2} - \dfrac{\pi}{3}$; $\theta = \dfrac{\pi}{6}$

3. $\theta + 7° = 90° - 48°$; $\theta = 35°$

4. $\theta - 40° = 90° - 63°$; $\theta = 67°$

5. $5\theta + \dfrac{\pi}{12} = \dfrac{\pi}{2} - \left(3\theta - \dfrac{\pi}{4}\right)$; $\theta = \dfrac{\pi}{12}$

EXERCISE 6-3

1. $\sin\theta = \dfrac{\sqrt{3}}{2}$, $\cos\theta = \dfrac{1}{2}$, $\tan\theta = \sqrt{3}$

2. $\sin\theta = -\dfrac{1}{2}$, $\cos\theta = \dfrac{\sqrt{3}}{2}$, $\tan\theta = -\dfrac{1}{\sqrt{3}}$

3. $\sin\theta = \dfrac{\sqrt{2}}{2}$, $\cos\theta = -\dfrac{\sqrt{2}}{2}$, $\tan\theta = -1$

4. $\sin\theta = -\dfrac{\sqrt{3}}{2}$, $\cos\theta = -\dfrac{1}{2}$, $\tan\theta = \sqrt{3}$

5. $\sin\theta = \dfrac{5}{13}$, $\cos\theta = -\dfrac{12}{13}$, $\tan\theta = -\dfrac{5}{12}$

6. $\sec\theta = -\dfrac{41}{9}$, $\csc\theta = -\dfrac{41}{40}$, $\cot\theta = \dfrac{9}{40}$

7. $\sec\theta = \dfrac{2}{\sqrt{2}}, \csc\theta = -\dfrac{2}{\sqrt{2}}, \cot\theta = -1$

8. $\sec\theta = -\dfrac{13}{12}, \csc\theta = \dfrac{13}{5}, \cot\theta = -\dfrac{12}{5}$

9. $\sec\theta = -\dfrac{25}{7}, \csc\theta = \dfrac{25}{24}, \cot\theta = -\dfrac{7}{24}$

10. $\sec\theta = \dfrac{85}{36}, \csc\theta = -\dfrac{85}{77}, \cot\theta = -\dfrac{36}{77}$

EXERCISE 6-4

1. 2

2. 6

3. 4

4. 0

5. 6

EXERCISE 6-5

1. $\cos 30° = \dfrac{\sqrt{3}}{2}$

2. $\sec 45° = \sqrt{2}$

3. $\sin\dfrac{\pi}{3} = \dfrac{\sqrt{3}}{2}$

4. $\csc 90° = 1$

5. $\tan\dfrac{5\pi}{4} = 1$

6. $5\sqrt{3}\tan\dfrac{\pi}{6} = 5\sqrt{3}\left(\dfrac{1}{\sqrt{3}}\right) = 5$

7. $6\sin 30° - 2\cos 60° = 2$

8. $3\sqrt{2}\sin 45° + 2\sqrt{3}\cos 30° = 3\sqrt{2}\left(\dfrac{\sqrt{2}}{2}\right) + 2\sqrt{3}\left(\dfrac{\sqrt{3}}{2}\right) = 6$

9. $-\tan\dfrac{\pi}{4}\sin\dfrac{\pi}{6} = -\dfrac{1}{2}$

10. $\cos\dfrac{\pi}{3}\sin\dfrac{\pi}{3} = \dfrac{\sqrt{3}}{4}$

EXERCISE 6-6

1. $\tan\theta = -\dfrac{1}{3}$; $\cot\theta = -3$

2. $\sin\theta = -\dfrac{11}{60}$; $\csc\theta = -\dfrac{60}{11}$

3. $\sec\theta = -\dfrac{25}{7}$; $\cos\theta = -\dfrac{7}{25}$

4. $\cot\theta = \dfrac{3}{4}$; $\tan\theta = \dfrac{4}{3}$

5. $\cos\theta = \dfrac{48}{73}$; $\sec\theta = \dfrac{73}{48}$

6. $\cos(-60°) = \cos 60° = \dfrac{1}{2}$

7. $\sin\left(-\dfrac{\pi}{6}\right) = -\sin\dfrac{\pi}{6} = -\dfrac{1}{2}$

8. $\sin(-30°) = -\sin 30° = -\dfrac{1}{2}$

9. $\cot\left(-\dfrac{\pi}{4}\right) = -\cot\dfrac{\pi}{4} = -1$

10. $-5\sqrt{2}\tan(45°) + 4\cos(60°) = -5\sqrt{2} + 2$

EXERCISE 6-7

1. 1

2. $-\dfrac{1}{\sqrt{3}}$

3. -1

4. $-\dfrac{1}{\sqrt{3}}$

5. $\dfrac{\sqrt{2}}{2}$

6. $-\sqrt{2}$

7. $-\sqrt{2}$

8. $2\sqrt{3}$

9. $\dfrac{\sqrt{2}}{4}$

10. $-\dfrac{1}{\sqrt{3}}$

Trigonometric Identities

EXERCISE 7-1

1. variable(s)

2. 0

3. identical

4. counterexample

5. cannot

EXERCISE 7-2

1. $\dfrac{\sin^3 \theta}{\cos^3 \theta} = \left(\dfrac{\sin \theta}{\cos \theta}\right)^3 = \tan^3 \theta$

2. $\dfrac{\sec \theta}{\tan \theta} = \dfrac{1}{\cancel{\cos \theta}} \cdot \dfrac{\cancel{\cos \theta}}{\sin \theta} = \dfrac{1}{\sin \theta} = \csc \theta$

3. $\sec \theta \cot \theta = \dfrac{1}{\cancel{\cos \theta}} \cdot \dfrac{\cancel{\cos \theta}}{\sin \theta} = \csc \theta$

4. $\csc \theta \tan \theta = \dfrac{1}{\cancel{\sin \theta}} \cdot \dfrac{\cancel{\sin \theta}}{\cos \theta} = \dfrac{1}{\cos \theta} = \sec \theta$

5. $\dfrac{\csc\theta}{\cot\theta} = \dfrac{1}{\sin\theta} \cdot \dfrac{\sin\theta}{\cos\theta} = \dfrac{1}{\cos\theta} = \sec\theta$

6. $\dfrac{\cot^2\theta}{\csc^2\theta} = \dfrac{\cos^2\theta}{\sin^2\theta} \cdot \dfrac{\sin^2\theta}{1} = \cos^2\theta$

7. $\dfrac{\sin^2 2\theta \cot 2\theta}{\cos 2\theta} = \dfrac{\sin^2 2\theta \cos 2\theta}{\cos 2\theta \sin 2\theta} = \sin 2\theta$

8. $\sin\theta\cot^2\theta\sec^2\theta = \sin\theta \dfrac{\cos^2\theta}{\sin^2\theta} \dfrac{1}{\cos^2\theta} = \dfrac{1}{\sin\theta} = \csc\theta$

9. $\dfrac{\sec\theta}{\cot\theta} = \dfrac{\dfrac{1}{\cos\theta}}{\dfrac{\cos\theta}{\sin\theta}} = \dfrac{1}{\cos\theta}\dfrac{\sin\theta}{\cos\theta} = \dfrac{\sin\theta}{\cos^2\theta}$

10. $\dfrac{\cot^2\theta}{\sec^2\theta} = \dfrac{\dfrac{\cos^2\theta}{\sin^2\theta}}{\dfrac{1}{\cos^2\theta}} = \dfrac{\cos^2\theta}{\sin^2\theta} \cdot \dfrac{\cos^2\theta}{1} = \dfrac{\cos^4\theta}{\sin^2\theta}$

EXERCISE 7-3

1. $49°$

2. $\dfrac{4\pi}{15}$

3. $15°$

4. $\dfrac{\pi}{10}$

5. $\dfrac{13\pi}{90}$

6. $\sin 30° = \dfrac{1}{2}$

7. $\sec\dfrac{\pi}{4} = \sqrt{2}$

8. $\tan 240° = \sqrt{3}$

9. $\dfrac{\sqrt{2}\cot\dfrac{\pi}{4}}{\sin\dfrac{7\pi}{4}} = \dfrac{\sqrt{2}}{-\dfrac{\sqrt{2}}{2}} = -2$

10. $\dfrac{3}{4}$

11. $-\sin(45°) - \cos(60°) = -\dfrac{\sqrt{2}}{2} - \dfrac{1}{2} = -\dfrac{\sqrt{2}+1}{2}$

12. $\dfrac{-\csc\left(\dfrac{\pi}{3}\right)}{-\sin\left(\dfrac{\pi}{2}\right)}\cos\left(\dfrac{\pi}{6}\right) = 1$

13. $-\sqrt{3}\tan\left(\dfrac{\pi}{3}\right) = -3$

14. $4\sin(90°)\left[-\csc(90°) - \sec(180°)\right] = 0$

15. $\sqrt{2}\sec(45°)\sin(30°) = 1$

EXERCISE 7-4

1. $\cot^2\theta + 1 = \dfrac{\cos^2\theta}{\sin^2\theta} + 1 = \dfrac{\cos^2\theta + \sin^2\theta}{\sin^2\theta} = \dfrac{1}{\sin^2\theta} = \csc^2\theta$

2. $\csc\theta + \sin\theta = \dfrac{1}{\sin\theta} + \sin\theta = \dfrac{\sin^2\theta + 1}{\sin\theta}$

3. $1 + \tan^2\theta = \sec^2\theta = \dfrac{1}{\cos^2\theta}$

4. $\cos^4\theta - \sin^4\theta = (\cos^2\theta - \sin^2\theta)(\cos^2\theta + \sin^2\theta) = \cos^2\theta - \sin^2\theta$

5. $\cos^2\theta(1 + \tan^2\theta) = \cos^2\theta\sec^2\theta = \cos^2\theta\left(\dfrac{1}{\cos^2\theta}\right) = 1$

6. $\dfrac{\sin^2\theta}{1-\cos\theta} = \dfrac{1-\cos^2\theta}{1-\cos\theta} = \dfrac{(1-\cos\theta)(1+\cos\theta)}{1-\cos\theta} = 1+\cos\theta$

$= 1 + \dfrac{1}{\sec\theta} = \dfrac{\sec\theta+1}{\sec\theta} = \dfrac{1+\sec\theta}{\sec\theta}$

7. $\dfrac{1}{1-\sin\theta} + \dfrac{1}{1+\sin\theta} = \dfrac{(1+\sin\theta)+(1-\sin\theta)}{1-\sin^2\theta} = \dfrac{2}{\cos^2\theta}$

8. $(\csc^2\theta - 1)(\sec^2\theta - 1) = \cot^2\theta\tan^2\theta = \dfrac{1}{\tan^2\theta}\tan^2\theta = 1$

9. $\dfrac{1+\tan^2\theta}{\csc^2\theta} = \dfrac{\sec^2\theta}{\csc^2\theta} = \dfrac{\dfrac{1}{\cos^2\theta}}{\dfrac{1}{\sin^2\theta}} = \dfrac{\sin^2\theta}{\cos^2\theta} = \left(\dfrac{\sin\theta}{\cos\theta}\right)^2 = \tan^2\theta$

10. $\dfrac{\sin^2\theta}{\tan^2\theta - \sin^2\theta} = \dfrac{\sin^2\theta}{\sin^2\theta\left(\dfrac{1}{\cos^2\theta}-1\right)} = \dfrac{\cos^2\theta}{1-\cos^2\theta} = \dfrac{\cos^2\theta}{\sin^2\theta} = \cot^2\theta$

EXERCISE 7-5

1. $\sin(60° + 45°) = \dfrac{\sqrt{6}+\sqrt{2}}{4}$

2. $\sin(30° + 45°) = \dfrac{\sqrt{6}+\sqrt{2}}{4}$

3. $\sin\left(\dfrac{\pi}{6} + \dfrac{\pi}{4}\right) = \dfrac{\sqrt{6}+\sqrt{2}}{4}$

4. $\sin\left(\dfrac{\pi}{4} + \dfrac{5\pi}{6}\right) = \dfrac{\sqrt{2}-\sqrt{6}}{4}$

5. $\dfrac{\sqrt{3}}{2}$

6. $\sin 6\theta$

7. $\sin 4\theta$

8. $\sin 2\theta$

9. $\sin \theta$

10. $\sin \theta$

11. $\sin(180° + \theta) = \sin 180° \cos \theta + \cos 180° \sin \theta$
$$= 0 \cdot \cos \theta + (-1)\sin \theta = -\sin \theta$$

12. $\sin(360° - \theta) = \sin 360° \cos \theta - \cos 360° \sin \theta$
$$= 0 \cdot \cos \theta - (1)\sin \theta = -\sin \theta$$

13. $\sin(90° - \theta) = \sin 90° \cos \theta - \cos 90° \sin \theta = 1 \cdot \cos \theta - (0) \cdot \sin \theta = \cos \theta$

14. $\sin\left(\dfrac{3\pi}{2} + \theta\right) = \sin\dfrac{3\pi}{2}\cos \theta + \cos\dfrac{3\pi}{2}\sin \theta = (-1)\cos \theta + (0)\sin \theta$
$$= -\cos\theta$$

15. $\sin\left(\theta - \dfrac{\pi}{6}\right) = \sin\theta\cos\left(\dfrac{\pi}{6}\right) - \cos\theta\sin\left(\dfrac{\pi}{6}\right) = \sin\theta\left(\dfrac{\sqrt{3}}{2}\right) - \cos\theta\left(\dfrac{1}{2}\right)$
$$= \dfrac{\sqrt{3}\sin\theta - \cos\theta}{2}$$

EXERCISE 7-6

1. $\cos(150° + 45°) = -\dfrac{\sqrt{2} + \sqrt{6}}{4}$

2. $\cos\left(\dfrac{\pi}{6} + \dfrac{\pi}{4}\right) = \dfrac{\sqrt{6} - \sqrt{2}}{4}$

3. $\cos(135° + 30°) = -\dfrac{\sqrt{6} + \sqrt{2}}{4}$

4. $\cos\left(\dfrac{\pi}{3} - \dfrac{\pi}{4}\right) = \dfrac{\sqrt{2} + \sqrt{6}}{4}$

5. $-\dfrac{1}{2}$

6. $\cos \theta$

7. $\cos 2\theta$

8. $\cos 4\theta$

9. $\cos 2\theta$

10. $\cos 2\theta$

11. $\cos(180° - \theta) = \cos 180° \cos \theta + \sin 180° \sin \theta = (-1)\cos \theta + (0)\sin \theta$
$$= -\cos \theta$$

12. $\cos(180° + \theta) = \cos 180° \cos \theta - \sin 180° \sin \theta = (-1)\cos \theta - (0)\sin \theta$
$$= -\cos \theta$$

13. $\cos(360° - \theta) = \cos 360° \cos \theta + \sin 360° \sin \theta = (1)\cos \theta + (0)\sin \theta$
$$= \cos \theta$$

14. $\cos\left(\dfrac{5\pi}{2} + \theta\right) = \cos\dfrac{5\pi}{2}\cos \theta - \sin\dfrac{5\pi}{2}\sin \theta = (0)\cos \theta - (1)\sin \theta = -\sin \theta$

15. $\cos\left(\theta - \dfrac{\pi}{3}\right) = \cos \theta \cos\dfrac{\pi}{3} + \sin \theta \sin\dfrac{\pi}{3} = \cos \theta\left(\dfrac{1}{2}\right) + \sin \theta\left(\dfrac{\sqrt{3}}{2}\right)$
$$= \dfrac{\cos \theta + \sqrt{3}\sin \theta}{2}$$

EXERCISE 7-7

1. $\tan(150° + 45°) = \dfrac{\sqrt{3} - 1}{\sqrt{3} + 1}$

2. $\tan(30° + 45°) = \dfrac{\sqrt{3} + 1}{\sqrt{3} - 1}$

3. $\tan(60° + 45°) = \dfrac{\sqrt{3} + 1}{1 - \sqrt{3}}$

4. $\tan\left(\dfrac{\pi}{3} - \dfrac{\pi}{4}\right) = \dfrac{\sqrt{3} - 1}{\sqrt{3} + 1}$

5. $\tan(135° + 30°) = \dfrac{1 - \sqrt{3}}{\sqrt{3} + 1}$

6. $\tan(130° + 50°) = \tan 180° = 0$

7. $\tan(110° - 50°) = \tan 60° = \sqrt{3}$

8. $\tan(115° - 70°) = \tan 45° = 1$

9. $\tan(100° + 50°) = \tan 150° = -\dfrac{1}{\sqrt{3}}$

10. $\tan(95° + 40°) = \tan 135° = -1$

11. $\tan 9\theta$

12. $\tan 4\theta$

13. $\tan 14\theta$

14. $\tan \theta$

15. $\tan 2\theta$

EXERCISE 7-8

1. $2\tan 4\theta$

2. $\cos 10\theta$

3. $\sin 4\theta$

4. $\tan 2\theta$

5. $\cos 6\theta$

6. $\sin 2\theta \csc \theta = \dfrac{2\sin\theta\cos\theta}{\sin\theta} = 2\cos\theta$

7. $\dfrac{2}{1+\cos 2\theta} = \dfrac{2}{1+2\cos^2\theta - 1} = \dfrac{2}{2\cos^2\theta} = \sec^2\theta$

8. $\dfrac{1-\tan^2 2\theta}{2\tan 2\theta} = \dfrac{1}{\dfrac{2\tan 2\theta}{1-\tan^2 2\theta}} = \dfrac{1}{\tan 4\theta} = \cot 4\theta$

9. $\dfrac{1}{2}\cot\theta + \dfrac{1}{2}\tan\theta = \dfrac{\cot\theta + \tan\theta}{2} = \dfrac{1}{2}\left[\dfrac{\cos\theta}{\sin\theta} + \dfrac{\sin\theta}{\cos\theta}\right]$

$= \dfrac{1}{2}\left[\dfrac{\cos^2\theta + \sin^2\theta}{\sin\theta\cos\theta}\right] = \dfrac{1}{2\sin\theta\cos\theta} = \dfrac{1}{\sin 2\theta} = \csc 2\theta$

10. $4\cos^3\theta\sin\theta - 4\cos\theta\sin^3\theta = 4\cos\theta\sin\theta\left[\cos^2\theta - \sin^2\theta\right]$

$= 2\sin 2\theta\cos 2\theta = \sin 4\theta$

EXERCISE 7-9

1. $\sin\left(\dfrac{1}{2}\cdot\dfrac{\pi}{6}\right)=\sqrt{\dfrac{1-\cos\dfrac{\pi}{6}}{2}}=\sqrt{\dfrac{1-\dfrac{\sqrt{3}}{2}}{2}}=\dfrac{\sqrt{2-\sqrt{3}}}{2}$

2. $\tan\left(\dfrac{30°}{2}\right)=\dfrac{\sin 30°}{1+\cos 30°}=\dfrac{\dfrac{1}{2}}{1+\dfrac{\sqrt{3}}{2}}=\dfrac{1}{2+\sqrt{3}}$

3. $\cos\left(\dfrac{1}{2}\cdot\dfrac{7\pi}{6}\right)=-\sqrt{\dfrac{1+\cos\dfrac{7\pi}{6}}{2}}=-\sqrt{\dfrac{1-\dfrac{\sqrt{3}}{2}}{2}}=-\dfrac{\sqrt{2-\sqrt{3}}}{2}$

4. $\cos\left(\dfrac{30°}{2}\right)=\sqrt{\dfrac{1+\cos 30°}{2}}=\sqrt{\dfrac{1+\dfrac{\sqrt{3}}{2}}{2}}=\dfrac{\sqrt{2+\sqrt{3}}}{2}$

5. $\cos\left(\dfrac{135°}{2}\right)=\sqrt{\dfrac{1+\cos 135°}{2}}=\sqrt{\dfrac{1-\dfrac{\sqrt{2}}{2}}{2}}=\dfrac{\sqrt{2-\sqrt{2}}}{2}$

8

Trigonometric Functions of Real Numbers

EXERCISE 8-1

1. $\sin\dfrac{5\pi}{3}=-\dfrac{\sqrt{3}}{2}$, $\cos\dfrac{5\pi}{3}=\dfrac{1}{2}$, $\tan\dfrac{5\pi}{3}=-\sqrt{3}$

2. $\sin\left(-\dfrac{3\pi}{4}\right)=-\dfrac{\sqrt{2}}{2}$, $\cos\left(-\dfrac{3\pi}{4}\right)=-\dfrac{\sqrt{2}}{2}$, $\tan\left(-\dfrac{3\pi}{4}\right)=1$

3. $\sin\dfrac{\pi}{6}=\dfrac{1}{2}$, $\cos\dfrac{\pi}{6}=\dfrac{\sqrt{3}}{2}$, $\tan\dfrac{\pi}{6}=\dfrac{1}{\sqrt{3}}$

4. $\sin\left(-\dfrac{11\pi}{6}\right) = \dfrac{1}{2}$, $\cos\left(-\dfrac{11\pi}{6}\right) = \dfrac{\sqrt{3}}{2}$, $\tan\left(-\dfrac{11\pi}{6}\right) = \dfrac{1}{\sqrt{3}}$

5. $\sin\pi = 0$, $\cos\pi = -1$, $\tan\pi = 0$

6. $\sec\dfrac{5\pi}{3} = 2$, $\csc\dfrac{5\pi}{3} = -\dfrac{2}{\sqrt{3}}$, $\cot\dfrac{5\pi}{3} = -\dfrac{1}{\sqrt{3}}$

7. $\sec\left(-\dfrac{3\pi}{4}\right) = -\sqrt{2}$, $\csc\left(-\dfrac{3\pi}{4}\right) = -\sqrt{2}$, $\cot\left(-\dfrac{3\pi}{4}\right) = 1$

8. $\sec\dfrac{\pi}{6} = \dfrac{2}{\sqrt{3}}$, $\csc\dfrac{\pi}{6} = 2$, $\cot\dfrac{\pi}{6} = \sqrt{3}$

9. $\sec\left(-\dfrac{11\pi}{6}\right) = \dfrac{2}{\sqrt{3}}$, $\csc\left(-\dfrac{11\pi}{6}\right) = 2$, $\cot\left(-\dfrac{11\pi}{6}\right) = \sqrt{3}$

10. $\sec\dfrac{2\pi}{3} = -2$, $\csc\dfrac{2\pi}{3} = \dfrac{2}{\sqrt{3}}$, $\cot\dfrac{2\pi}{3} = -\dfrac{1}{\sqrt{3}}$

11. 0.841

12. 0.408

13. −1.470

14. −1.010

15. 0.323

16. −0.182

17. 0.608

18. −0.432

19. 1.139

20. 1.662

EXERCISE 8-2

1. d

2. b

3. b

4. b

5. d

Graphs of the Sine Function

EXERCISE 9-1

1. $0, \pi, 2\pi$

2. $\dfrac{\pi}{2}$

3. $\dfrac{3\pi}{2}$

4. positive for $0 < x < \pi$

5. negative for $\pi < x < 2\pi$

6. Increasing for $0 < x < \dfrac{\pi}{2}$ and for $\dfrac{3\pi}{2} < x < 2\pi$

7. Decreasing for $\dfrac{\pi}{2} < x < \dfrac{3\pi}{2}$

8. Yes, because the domain of the sine function is all real numbers.

9. No, because the range of the sine function is $-1 \le y \le 1$.

10. Yes, because the sine function is an odd function.

EXERCISE 9-2

1. amplitude: 4, range: $-4 \le y \le 4$

2. amplitude: $\dfrac{1}{3}$, range: $-\dfrac{1}{3} \le y \le \dfrac{1}{3}$

3. amplitude: 1.5, range: $-1.5 \le y \le 1.5$

4. amplitude: 5, range: $-5 \leq y \leq 5$

5. amplitude: $\sqrt{2}$, range: $-\sqrt{2} \leq y \leq \sqrt{2}$

6. 10

7. $-\dfrac{4}{5}$

8. $0, \pi, 2\pi$

9. $\dfrac{3\pi}{2}$

10. $\dfrac{\pi}{2}$

11.

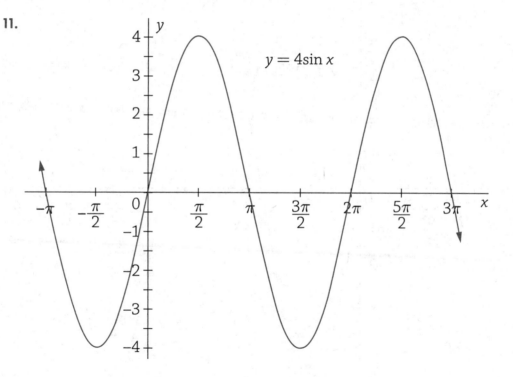

$y = 4\sin x$

12.

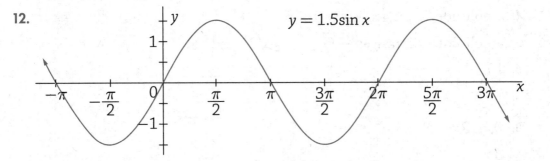

$y = 1.5\sin x$

13.

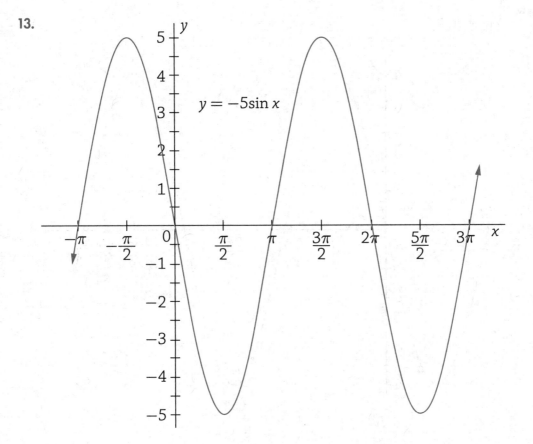

$y = -5\sin x$

14.

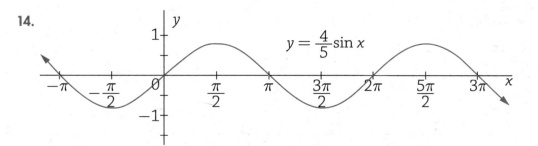

$y = \dfrac{4}{5}\sin x$

15.

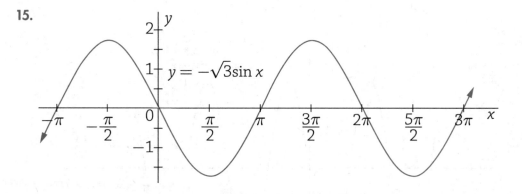

$y = -\sqrt{3}\sin x$

EXERCISE 9-3

1. $\dfrac{\pi}{2}$

2. $\dfrac{\pi}{3}$

3. 6π

4. 4π

5. 2

6. $0, \dfrac{\pi}{4}, \dfrac{\pi}{2}, \dfrac{3\pi}{4}, \pi, \dfrac{5\pi}{4}, \dfrac{3\pi}{2}, \dfrac{7\pi}{4}, 2\pi$

7. $\dfrac{\pi}{8}, \dfrac{5\pi}{8}, \dfrac{9\pi}{8}, \dfrac{13\pi}{8}$

8. $\dfrac{3\pi}{8}, \dfrac{7\pi}{8}, \dfrac{11\pi}{8}, \dfrac{15\pi}{8}$

9. $0, 2\pi, 4\pi$

10. 3π

11.

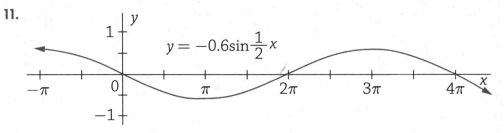

$y = -0.6\sin\frac{1}{2}x$

12.

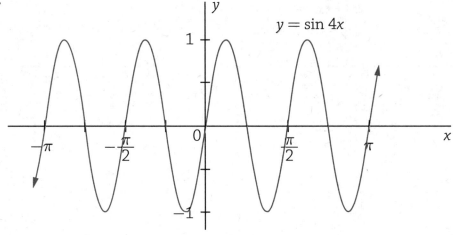

$y = \sin 4x$

13.

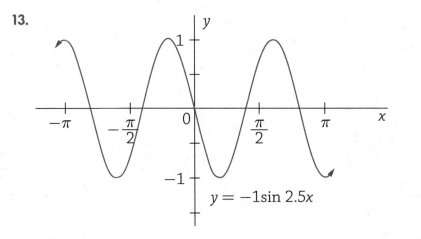

$y = -1\sin 2.5x$

14.

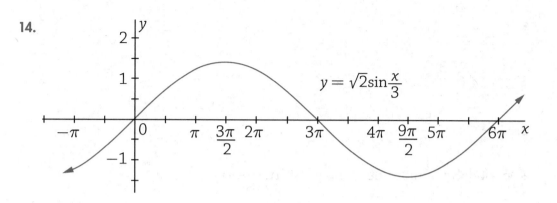

$y = \sqrt{2}\sin\dfrac{x}{3}$

15.

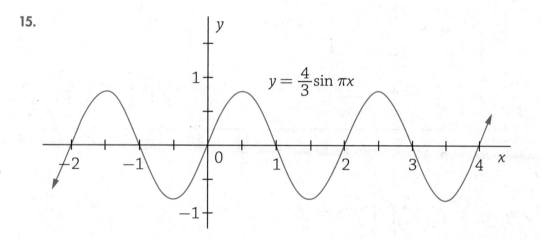

$y = \dfrac{4}{3}\sin \pi x$

EXERCISE 9-4

1. $\dfrac{3\pi}{8} \leq x \leq \dfrac{27\pi}{8}$

2. $\dfrac{\pi}{4} \leq x \leq \dfrac{5\pi}{4}$

3. $2 \leq x \leq 2 + \pi$

4. $\dfrac{1}{2} \leq x \leq \dfrac{3}{2}$

5. $-\dfrac{\pi}{3} \leq x \leq \dfrac{5\pi}{3}$

6. period: 3π; amplitude: 3; phase shift: $\dfrac{3\pi}{8}$, right

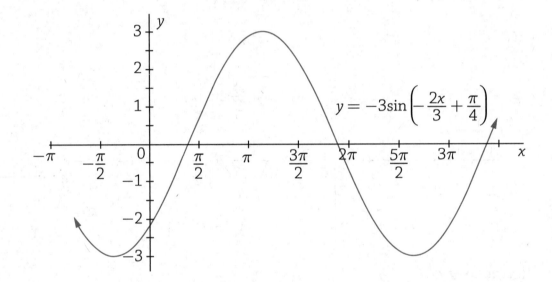

$$y = -3\sin\left(-\dfrac{2x}{3} + \dfrac{\pi}{4}\right)$$

7. period: π; amplitude: 2; phase shift: $\dfrac{\pi}{4}$, right

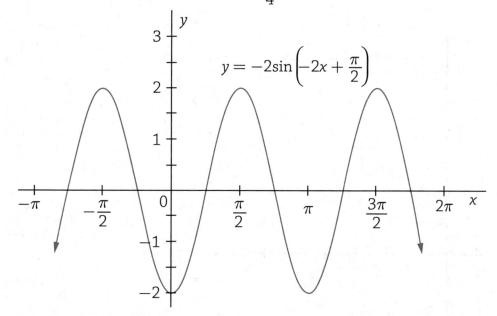

$$y = -2\sin\left(-2x + \dfrac{\pi}{2}\right)$$

8. period: π; amplitude: $\dfrac{1}{2}$; phase shift: 2, right; reflection over the x-axis

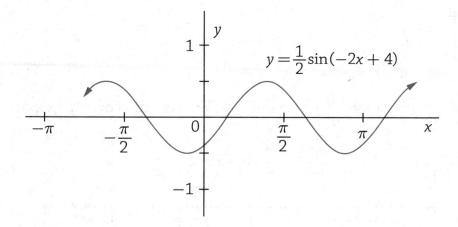

$$y = \dfrac{1}{2}\sin(-2x + 4)$$

9. period: 1; amplitude: 3; phase shift: $\dfrac{1}{2}$, right; reflection over the x-axis

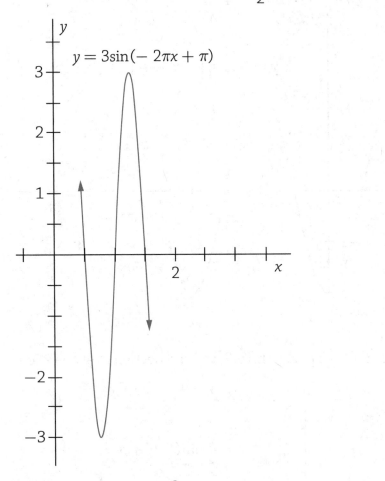

$$y = 3\sin(-2\pi x + \pi)$$

10. period: 2π; amplitude: $\dfrac{3}{4}$; phase shift: $\dfrac{\pi}{3}$, left; reflection over the x-axis

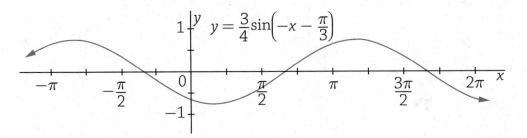

$$y = \dfrac{3}{4}\sin\left(-x - \dfrac{\pi}{3}\right)$$

EXERCISE 9-5

1. period: 2π; amplitude: 5; phase shift: $\dfrac{\pi}{4}$, right; vertical shift: 1, down

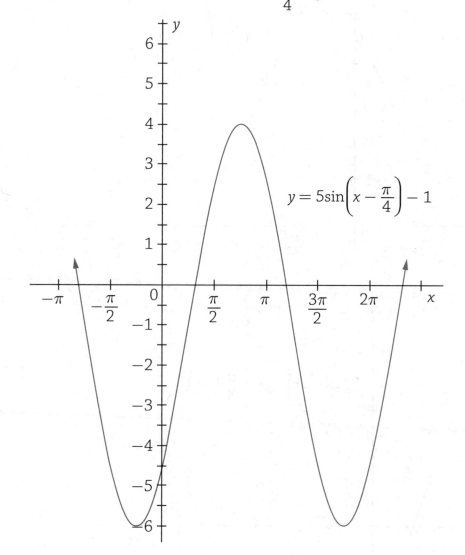

$$y = 5\sin\left(x - \frac{\pi}{4}\right) - 1$$

2. period: 2π; amplitude: 2; phase shift: $\dfrac{\pi}{4}$, left; vertical shift: 3, up

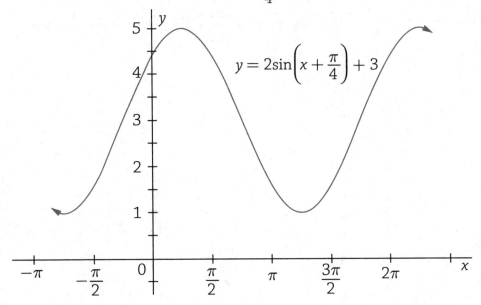

$$y = 2\sin\left(x + \dfrac{\pi}{4}\right) + 3$$

3. period: π; amplitude: 1; phase shift: $\dfrac{\pi}{6}$, right; vertical shift: $\dfrac{1}{2}$, up

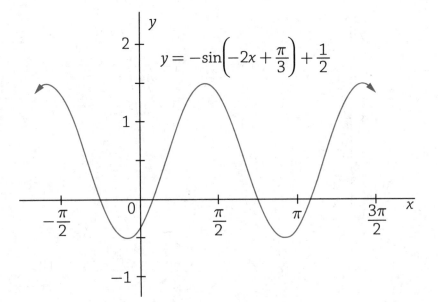

$$y = -\sin\left(-2x + \dfrac{\pi}{3}\right) + \dfrac{1}{2}$$

4. period: $\dfrac{2\pi}{3}$; amplitude: $\dfrac{1}{2}$; phase shift: $\dfrac{\pi}{3}$, left; vertical shift: 3, up; reflection over $y = 3$

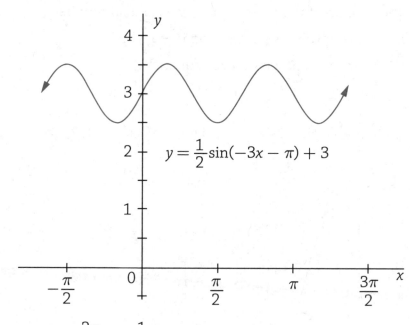

$$y = \frac{1}{2}\sin(-3x - \pi) + 3$$

5. period: $\dfrac{2\pi}{120\pi} = \dfrac{1}{60}$; maximum value: 15 amperes

Graphs of the Cosine Function

EXERCISE 10-1

1. $\dfrac{\pi}{2}, \dfrac{3\pi}{2}$

2. $0, 2\pi$

3. π

4. positive for $0 \le x < \dfrac{\pi}{2}$ and $\dfrac{3\pi}{2} < x \le 2\pi$

5. negative for $\dfrac{\pi}{2} < x < \dfrac{3\pi}{2}$

6. Increasing for $\pi < x < 2\pi$

7. Decreasing for $0 < x < \pi$

8. Yes, because the domain of the cosine function is all real numbers.

9. No, because the range of the cosine function is $-1 \le y \le 1$.

10. No, because the cosine function is an even function.

EXERCISE 10-2

1. period: 2π; amplitude: 5; phase shift: $\dfrac{\pi}{4}$, right; vertical shift: 1, down

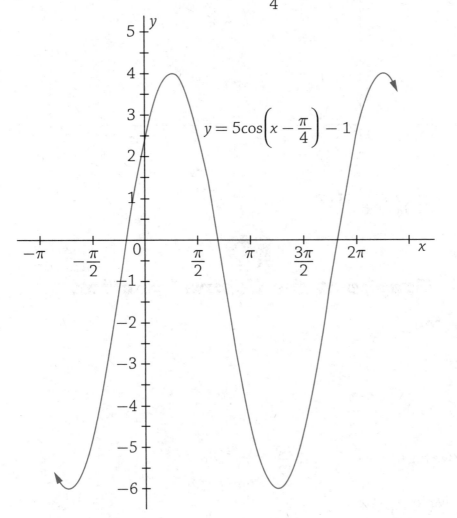

$$y = 5\cos\left(x - \frac{\pi}{4}\right) - 1$$

2. period: 2π; amplitude: 2; phase shift: $\dfrac{\pi}{4}$, left; vertical shift: 3, up

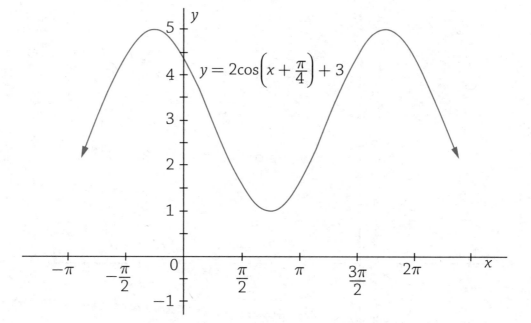

$$y = 2\cos\left(x + \dfrac{\pi}{4}\right) + 3$$

3. period: π; amplitude: 1; phase shift: $\dfrac{\pi}{6}$, right; vertical shift: $\dfrac{1}{2}$, up; reflection over $y = \dfrac{1}{2}$

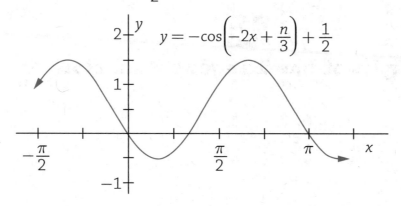

$$y = -\cos\left(-2x + \dfrac{n}{3}\right) + \dfrac{1}{2}$$

4. period: $\dfrac{2\pi}{3}$; amplitude: $\dfrac{1}{2}$; phase shift: $\dfrac{\pi}{3}$, left; vertical shift: 3, up

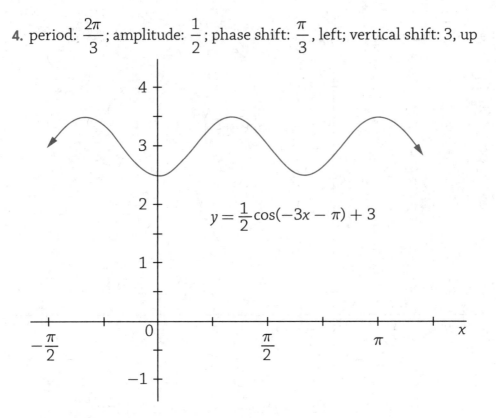

$$y = \frac{1}{2}\cos(-3x - \pi) + 3$$

5. period: $\dfrac{2\pi}{\pi} = 2$; maximum value: 10

11

Graphs of the Tangent Function

EXERCISE 11-1

1. R

2. $\dfrac{\pi}{2} + n\pi$

3. $n\pi$

4. π

5. undefined

EXERCISE 11-2

1. period: π; stretching factor: 3; horizontal shift: $\dfrac{\pi}{4}$, right; vertical shift: 2, down; asymptotes: $\dfrac{\pi}{4} + n\dfrac{\pi}{2}$, n odd

2. period: π; stretching factor: 5; horizontal shift: $\dfrac{\pi}{4}$, left; vertical shift: 1, up; asymptotes: $-\dfrac{\pi}{4} + n\dfrac{\pi}{2}$, n odd; reflection over $y = 1$

3. period: $\dfrac{\pi}{2}$; stretching factor: 4; horizontal shift: $\dfrac{\pi}{12}$, left; vertical shift: $\dfrac{1}{2}$, up; asymptotes: $-\dfrac{\pi}{12} + n\dfrac{\pi}{4}$, n odd

4. period: $\dfrac{\pi}{2}$; compression factor: $\dfrac{1}{2}$; horizontal shift: $\dfrac{\pi}{2}$, right; vertical shift: $\sqrt{5}$, up; asymptotes: $\dfrac{\pi}{2} + n\dfrac{\pi}{4}$, n odd

5. period: π; stretching factor: 1; horizontal shift: $\dfrac{\pi}{3}$, right; vertical shift: none; asymptotes: $\dfrac{\pi}{3} + n\dfrac{\pi}{2}$, n odd

$$y = -\tan\left(-x + \dfrac{\pi}{3}\right)$$

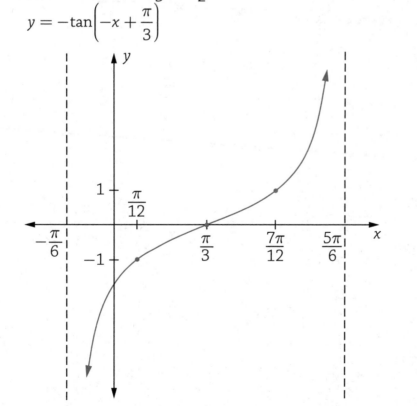

Inverse Trigonometric Functions

EXERCISE 12-1

1. F

2. T

3. T

4. T

5. F

EXERCISE 12-2

1. $\dfrac{\pi}{2}$

2. $-\dfrac{\pi}{4}$

3. $\dfrac{\pi}{6}$

4. $-\dfrac{\pi}{6}$

5. $\dfrac{5\pi}{6}$

6. $\sin\left(\sin^{-1}\dfrac{\sqrt{3}}{2}\right) = \sin\dfrac{\pi}{3} = \dfrac{\sqrt{3}}{2}$

7. $\tan\left(\cos^{-1}\dfrac{1}{2}\right) = \tan\dfrac{\pi}{3} = \sqrt{3}$

8. $\sin^{-1}\left(\sin\dfrac{3\pi}{4}\right) = \sin^{-1}\left(\dfrac{\sqrt{2}}{2}\right) = \dfrac{\pi}{4}$

9. $\cos^{-1}\left(\sin\dfrac{\pi}{6}\right) = \cos^{-1}\left(\dfrac{1}{2}\right) = \dfrac{\pi}{3}$

10. $\cos\left(\tan^{-1}\sqrt{3}\right) = \cos\left(\dfrac{\pi}{3}\right) = \dfrac{1}{2}$

11. $\dfrac{7}{25}$

12. $\dfrac{12}{13}$

13. $\dfrac{4}{3}$

14. $\dfrac{4}{x}$

15. $\dfrac{3}{x}$

16. 0.360

17. 0.836

18. −1.072

19. −1.388

20. 0.460

13
Solving Trigonometric Equations

EXERCISE 13-1

1. C

2. I

3. C

4. I

5. C

EXERCISE 13-2

1. $2\cos x + 1 = 0$

$$\cos x = -\frac{1}{2}$$

$$x = \frac{2\pi}{3} + n \cdot 2\pi \text{ or } x = \frac{4\pi}{3} + n \cdot 2\pi$$

2. $\tan\theta = -\sqrt{3}$

$$\theta = 120° + n \cdot 360° \text{ or } \theta = 300° + n \cdot 360°$$

3. $\cos^2 x - \sin^2 x = \dfrac{\sqrt{3}}{2}$

$$\cos 2x = \frac{\sqrt{3}}{2}$$

$$2x = \frac{\pi}{6} \text{ or } 2x = \frac{11\pi}{6}$$

$$x = \frac{\pi}{12} \text{ or } x = \frac{11\pi}{12}$$

4.
$$\sin^2 x \cos x = \frac{1}{4}\cos x$$

$$\sin^2 x \cos x - \frac{1}{4}\cos x = 0$$

$$\cos x\left(\sin^2 x - \frac{1}{4}\right) = 0$$

$$\cos x\left(\sin x + \frac{1}{2}\right)\left(\sin x - \frac{1}{2}\right) = 0$$

$$\cos x = 0, \sin x = -\frac{1}{2}, \sin x = \frac{1}{2}$$

$$x = \frac{\pi}{6}, \frac{\pi}{2}, \frac{5\pi}{6}, \frac{7\pi}{6}, \frac{3\pi}{2}, \text{or } \frac{11\pi}{6}$$

5. $\cos 3\theta = \dfrac{1}{2}$

$3\theta = 60° \text{ or } 3\theta = 300°$

$\theta = 20° \text{ or } \theta = 100°$

EXERCISE 13-3

1. $x \approx 0.244 + n \cdot 2\pi \text{ or } x \approx 2.897 + n \cdot 2\pi$

2. $\theta \approx 59.0° \text{ or } \theta \approx 301.0°$

3. $\theta \approx 155.3° \text{ or } \theta \approx 204.7°$

4. $x \approx 2.251 \text{ or } x \approx 2.461$

5. $\cos 2\theta = \dfrac{\left(9.81\frac{m}{s^2}\right)h}{v^2 + \left(9.81\frac{m}{s^2}\right)h} = \dfrac{\left(9.81\frac{m}{s^2}\right)(1.8\,m)}{\left(13.8\frac{m}{s}\right)^2 + \left(9.81\frac{m}{s^2}\right)(1.8\,m)} \approx 0.08485$

$2\theta \approx 85.13°;\ \theta \approx 42.6°$

Calculator Instructions for Trigonometry Using the TI-84 Plus

Because you are a serious student of mathematics, we assume you already have a graphing calculator and that you have mastered the calculator's elementary operation. In this appendix, you will see that a graphing calculator is an indispensable tool for working with trigonometry. To demonstrate basic trigonometry features of graphing calculators, we have elected to use the TI-84 Plus Silver Edition platform. If you have a different calculator, that's okay. Most graphing calculators will have trigonometry features like the ones shown here. Consult your user's guidebook for instructions.

General Usage

- Use the cursor arrow keys in the upper-right corner of the keypad to move the cursor around the screen.

- Use the ENTER key to evaluate expressions and to execute commands.

- Use the blue 2ND key to access the secondary options printed in blue above the keys.

BTW

For a 2ND key action, rather than showing the primary key function, this appendix shows the secondary option in brackets. For example, 2ND ∧ is shown as 2ND [π].

- Use [2ND] [QUIT] to exit a menu.

- Use [2ND] [ANS] to recall the previous result.

BTW

Do not press [CLEAR] to exit a menu because that action will sometimes delete your selection.

Setting the Calculator to Degree or Radian Mode

- To set your calculator to radian mode, press [MODE], scroll down to the third row, highlight RADIAN, and then press [ENTER].

```
NORMAL   SCI   ENG
FLOAT    0123456789
RADIAN   DEGREE
FUNC  PAR  POL  SEQ
CONNECTED   DOT
SEQUENTIAL   SIMUL
REAL  a+bi  re^θi
FULL  HORIZ  G-T
SET CLOCK 01/01/01 1:57AM
```

- To set your calculator to degree mode, press [MODE], scroll down to the third row, use the right arrow key to highlight DEGREE, and then press [ENTER].

```
NORMAL   SCI   ENG
FLOAT    0123456789
RADIAN   DEGREE
FUNC  PAR  POL  SEQ
CONNECTED   DOT
SEQUENTIAL   SIMUL
REAL  a+bi  re^θi
FULL  HORIZ  G-T
         01/01/01 2:03AM
```

Overriding Radian or Degree Mode

- To override radian mode, use 2ND [ANGLE] 1 : ° to enter a degree symbol after the entry and treat the result as a degree measure.

- To override degree mode, use 2ND [ANGLE] 3 : ᴿ to enter a radian symbol after the entry and treat the result as a radian measure.

Evaluating Trigonometric Functions

- Set the calculator to the desired mode (radians or degrees). To evaluate sine, cosine, or tangent, use their respective keypad keys.

EXAMPLE

▶ To evaluate $\sin \dfrac{\pi}{6}$, enter the keystroke sequence SIN 2ND [π] ÷ 6) ENTER in radian mode. (For convenience, numerical values are shown entered in full, rather than as separate keystrokes.)

▶ The display shows the following:

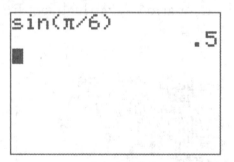

```
sin(π/6)
              .5
■
```

▶ Be sure to enclose fractional angle measures in parentheses. Notice that the calculator automatically inserts a left parenthesis.

EXAMPLE

▶ To evaluate $\cos 45°$, enter the keystroke sequence $\boxed{\text{COS}}$ 45 $\boxed{\text{ENTER}}$ in degree mode.

▶ The display shows the following:

```
cos(45
        .7071067812
█
```

▶ Notice that the closing parenthesis can be omitted when the function is at the end of a keystroke sequence.

EXAMPLE

▶ To evaluate the tangent of the real number −1,000, enter the keystroke sequence $\boxed{\text{TAN}}$ $\boxed{(-)}$ 1000 $\boxed{\text{ENTER}}$ in radian mode.

▶ The display shows the following:

```
tan(-1000
        -1.470324156
█
```

■ The TI-84 Plus (like most calculators) does not have built-in secant, cosecant, and cotangent functions. Use the $\boxed{x^{-1}}$ key with the associated reciprocal identities to evaluate these functions.

BTW

Be sure to use $\boxed{(-)}$, the negative key, not $\boxed{-}$, the minus key, to enter a negative value.

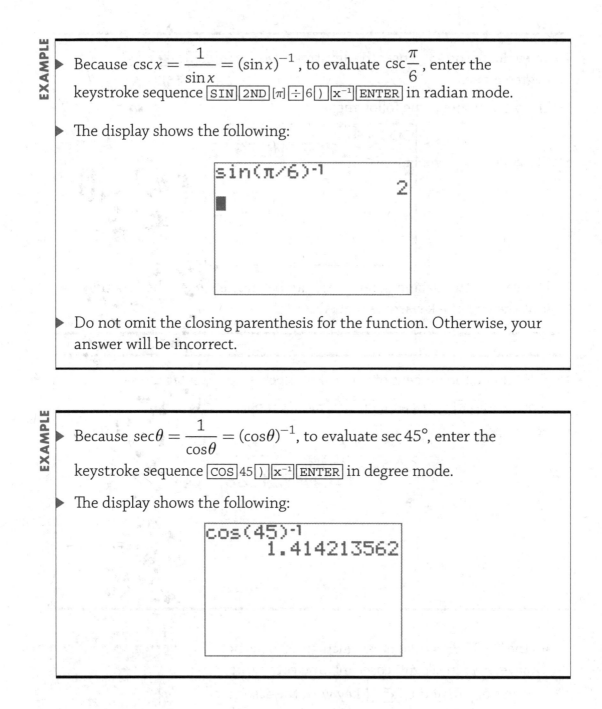

▶ Because $\csc x = \dfrac{1}{\sin x} = (\sin x)^{-1}$, to evaluate $\csc \dfrac{\pi}{6}$, enter the keystroke sequence $\boxed{\text{SIN}}\,\boxed{\text{2ND}}\,[\pi]\,\boxed{\div}\,6\,\boxed{)}\,\boxed{x^{-1}}\,\boxed{\text{ENTER}}$ in radian mode.

▶ The display shows the following:

```
sin(π/6)⁻¹
                    2
■
```

▶ Do not omit the closing parenthesis for the function. Otherwise, your answer will be incorrect.

▶ Because $\sec \theta = \dfrac{1}{\cos \theta} = (\cos \theta)^{-1}$, to evaluate $\sec 45°$, enter the keystroke sequence $\boxed{\text{COS}}\,45\,\boxed{)}\,\boxed{x^{-1}}\,\boxed{\text{ENTER}}$ in degree mode.

▶ The display shows the following:

```
cos(45)⁻¹
          1.414213562
```

EXAMPLE

▶ Because $\cot x = \dfrac{1}{\tan x} = (\tan x)^{-1}$, to evaluate the cotangent of the real number $-1{,}000$, enter the keystroke sequence $\boxed{\text{TAN}}\boxed{(-)}1000\boxed{)}\boxed{x^{-1}}$ $\boxed{\text{ENTER}}$ in radian mode.

▶ The display shows the following:

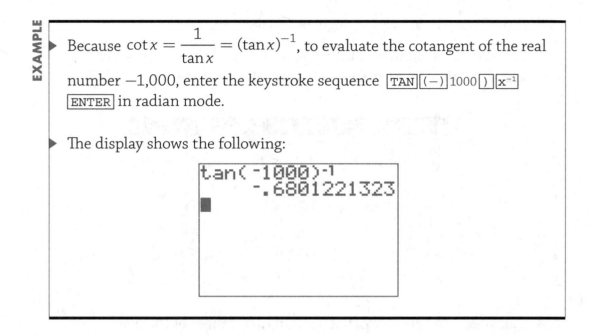

Determining Inverse Sine, Inverse Cosine, and Inverse Tangent Trigonometric Values

- Set the calculator to the desired mode.

- If you want the answer in radians, use radian mode.

If you want the answer in radians expressed in terms of π, use degree mode to obtain the answer in degrees, then multiply by $\dfrac{\pi}{180}$ to convert degrees to radians.

- If you want the answer in degrees, use degree mode.

- Recall that the domains and ranges of the $y = \sin^{-1} x$, $y = \cos^{-1} x$, and $y = \tan^{-1} x$ are as shown in the following table:

Inverse Function	Domain	Range
$y = \sin^{-1} x$	$-1 \le x \le 1$	$-\dfrac{\pi}{2} \le y \le \dfrac{\pi}{2}$
$y = \cos^{-1} x$	$-1 \le x \le 1$	$0 \le y \le \pi$
$y = \tan^{-1} x$	$-\infty < x < \infty$	$-\dfrac{\pi}{2} < y < \dfrac{\pi}{2}$

- Calculators use the restricted domains and ranges of the inverse trigonometric functions. Therefore, for a given number, the calculator can return only one radian (or degree) value. Depending on the sign, $+$ or $-$, of the input number, the included quadrants in which an output may lie are quadrants I, II, or IV (QI, QII, or QIV). See the following table for a summary of the included quadrants.

Inverse Function Key	+ Input	− Input
$\sin^{-1}$	QI	QIV
$\cos^{-1}$	QI	QII
$\tan^{-1}$	QI	QIV

- To determine values of the inverse sine, inverse cosine, and inverse tangent functions, use their respective secondary function keys. For $\boxed{\text{sin}^{-1}}$ use $\boxed{\text{2ND}}$ $[\sin^{-1}]$, for $\cos^{-1}$ use $\boxed{\text{2ND}}$ $[\cos^{-1}]$, and for $\tan^{-1}$ use $\boxed{\text{2ND}}$ $[\tan^{-1}]$.

EXAMPLE

▶ To determine $\sin^{-1}(0.5)$ in degrees, enter the keystroke sequence $\boxed{\text{2ND}}\,[\text{sin}^{-1}]\,.5\,\boxed{\text{ENTER}}$ in degree mode.

▶ The display shows the following:

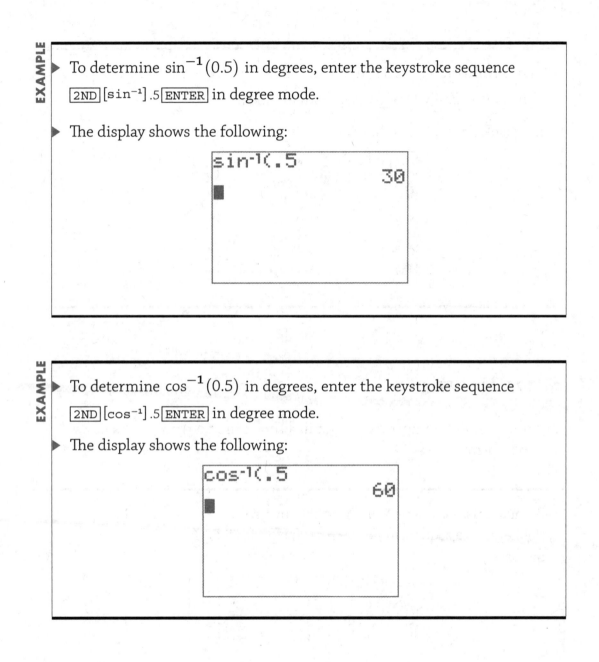

```
sin-1(.5
                    30
■
```

EXAMPLE

▶ To determine $\cos^{-1}(0.5)$ in degrees, enter the keystroke sequence $\boxed{\text{2ND}}\,[\text{cos}^{-1}]\,.5\,\boxed{\text{ENTER}}$ in degree mode.

▶ The display shows the following:

```
cos-1(.5
                    60
■
```

EXAMPLE

▶ To determine $\tan^{-1}\left(\sqrt{3}\right)$ in radians, enter the keystroke sequence
[2ND] [tan⁻¹] [2ND] [√] 3 [ENTER] in radian mode.

▶ The display shows the following:

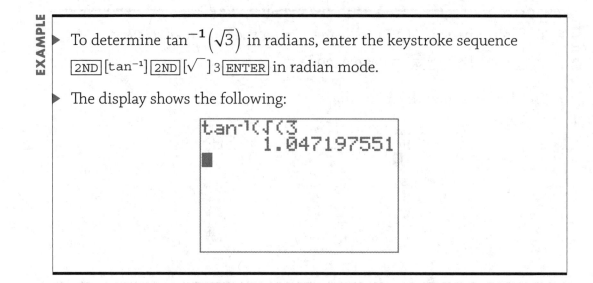

```
tan⁻¹(√(3
          1.047197551
▮
```

Error Messages

Entering an input number that is not in a function's domain results in a calculator error message.

EXAMPLE

▶ Because 5 is not in the domain of the inverse sine function, entering the keystroke sequence [2ND] [sin⁻¹] 5 [ENTER] yields the following message:

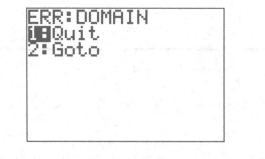

```
ERR:DOMAIN
1▮Quit
2:Goto
```

APPENDIX B

Trigonometric Identities/Formulas

Reciprocal Identities	Cofunction Identities	Periodic Identities
$\csc\theta = \dfrac{1}{\sin\theta}$	$\sin\left(\dfrac{\pi}{2} - \theta\right) = \cos\theta$	$\sin(\theta \pm n \cdot 2\pi) = \sin\theta$
$\sec\theta = \dfrac{1}{\cos\theta}$	$\cos\left(\dfrac{\pi}{2} - \theta\right) = \sin\theta$	$\cos(\theta \pm n \cdot 2\pi) = \cos\theta$
$\cot\theta = \dfrac{1}{\tan\theta}$	$\tan\left(\dfrac{\pi}{2} - \theta\right) = \cot\theta$	$\tan(\theta \pm n \cdot \pi) = \tan\theta$

Ratio Identities	Pythagorean Identities	Addition/Subtraction Formulas
$\tan\theta = \dfrac{\sin\theta}{\cos\theta}$	$\sin^2\theta + \cos^2\theta = 1$	$\sin(\theta \pm \varphi) = \sin\theta\cos\varphi \pm \cos\theta\sin\varphi$
$\cot\theta = \dfrac{\cos\theta}{\sin\theta}$	$\tan^2\theta + 1 = \sec^2\theta$	$\cos(\theta \pm \varphi) = \cos\theta\cos\varphi \mp \sin\theta\sin\varphi$
	$\cot^2\theta + 1 = \csc^2\theta$	$\tan(\theta \pm \varphi) = \dfrac{\tan\theta \pm \tan\varphi}{1 \mp \tan\theta\tan\varphi}$

Odd/Even Identities	Double-Angle Identities	Half-Angle Identities
$\sin(-\theta) = -\sin\theta$	$\sin 2\theta = 2\sin\theta\cos\theta$	$\sin\dfrac{\theta}{2} = \pm\sqrt{\dfrac{1-\cos\theta}{2}}$
$\cos(-\theta) = \cos\theta$	$\cos 2\theta = \cos^2\theta - \sin^2\theta$	$\cos\dfrac{\theta}{2} = \pm\sqrt{\dfrac{1+\cos\theta}{2}}$
$\tan(-\theta) = -\tan\theta$	$\cos 2\theta = 2\cos^2\theta - 1$	$\tan\dfrac{\theta}{2} = \pm\sqrt{\dfrac{1-\cos\theta}{1+\cos\theta}}$
	$\cos 2\theta = 1 - 2\sin^2\theta$	$\tan\dfrac{\theta}{2} = \dfrac{\sin\theta}{1+\cos\theta}$
	$\tan 2\theta = \dfrac{2\tan\theta}{1-\tan^2\theta}$	$\tan\dfrac{\theta}{2} = \dfrac{1-\cos\theta}{\sin\theta}$

Product-to-Sum Formulas	Sum-to-Product Formulas
$\sin\theta\cos\varphi = \dfrac{\sin(\theta+\varphi) + \sin(\theta-\varphi)}{2}$	$\sin\theta + \sin\varphi = 2\sin\left(\dfrac{\theta+\varphi}{2}\right)\cos\left(\dfrac{\theta-\varphi}{2}\right)$
$\sin\theta\sin\varphi = -\dfrac{\cos(\theta+\varphi) - \cos(\theta-\varphi)}{2}$	$\sin\theta - \sin\varphi = 2\cos\left(\dfrac{\theta+\varphi}{2}\right)\sin\left(\dfrac{\theta-\varphi}{2}\right)$
$\cos\theta\sin\varphi = \dfrac{\sin(\theta+\varphi) - \sin(\theta-\varphi)}{2}$	$\cos\theta - \cos\varphi = -2\sin\left(\dfrac{\theta+\varphi}{2}\right)\sin\left(\dfrac{\theta-\varphi}{2}\right)$
$\cos\theta\cos\varphi = \dfrac{\cos(\theta+\varphi) + \cos(\theta-\varphi)}{2}$	$\cos\theta + \cos\varphi = 2\cos\left(\dfrac{\theta+\varphi}{2}\right)\cos\left(\dfrac{\theta-\varphi}{2}\right)$

NOTES

NOTES

NOTES

NOTES

NOTES